**THE IOWA STATE UNIVERSITY PRESS SERIES
IN THE HISTORY OF TECHNOLOGY AND SCIENCE**

**HAMILTON CRAVENS** / *Series Editor*

**IOWA STATE UNIVERSITY PRESS SERIES
IN THE HISTORY OF TECHNOLOGY AND SCIENCE**

Christiaan Huygens' *The Pendulum Clock
or Geometrical Demonstrations Concerning
the Motion of Pendula as Applied to Clocks*
Translated with Notes by Richard J. Blackwell
Introduction by H. J. M. Bos

The History of Modern Science:
A Guide to the Second Scientific Revolution, 1800–1950
by Stephen G. Brush

John B. Jervis:
An American Engineering Pioneer
by F. Daniel Larkin

Science with Practice: Charles E. Bessey and the
Maturing of American Botany
by Richard A. Overfield

**SCIENCE WITH PRACTICE**

# SCIENCE WITH PRACTICE

## CHARLES E. BESSEY and the Maturing of American Botany

**RICHARD A. OVERFIELD**

 IOWA STATE UNIVERSITY PRESS / AMES

To Mike, Nellie, Corky,
Julie, Doug

Richard A. Overfield is Professor of History, University of Nebraska at Omaha.

All photographs are courtesy of the University of Nebraska–Lincoln Archives.

© 1993 Iowa State University Press, Ames, Iowa 50010
All rights reserved, except as noted in the preface

Authorization to photocopy items for internal or personal use, or the internal or personal use of specific clients, is granted by Iowa State University Press, provided that the base fee of $.10 per copy is paid directly to the Copyright Clearance Center, 27 Congress Street, Salem, MA 01970. For those organizations that have been granted a photocopy license by CCC, a separate system of payment has been arranged. The fee code for users of the Transactional Reporting Service is 0-8138-1822-2/93 $.10.

⊚ Printed on acid-free paper in the United States of America

First edition, 1993

Library of Congress Cataloging-in-Publication Data
Overfield, Richard A.
    Science with practice : Charles E. Bessey and the maturing of American botany / Richard A. Overfield.—1st ed.
    p.   cm.—(Iowa State University Press series in the history of technology and science)
    Includes bibliographical references and index.
    ISBN 0-8138-1822-2 (alk. paper)
    1. Bessey, Charles E. (Charles Edwin), 1845–1915. 2. Botanists—United States—Biography. 3. Botany—United States—History.
I. Title.    II. Series.
QK31.B516084    1993
581′.092—dc20
[B]    91-16180

# CONTENTS

Introduction, ix
Preface, xi

1. **Laying the Foundations: The Iowa Years**, 3

2. **A Time of Promise but Uncertainty**, 23

3. **New Foundations: The Move to Nebraska**, 47

4. **Seeking Maturity: The "New Botany,"** 72

5. **Professionalization and Reform**, 100

6. **Grasslands and Forests**, 131

7. **The Scientist as Progressive:
   Reforming American Society**, 157

8. **Evolution and Classification:
   The Bessey System**, 179

Photo Section follows page 80
Notes, 201
Bibliography, 235
Index, 255

# INTRODUCTION

RICHARD A. OVERFIELD'S *Science with Practice: Charles E. Bessey and the Maturing of American Botany* is the fourth volume published in the Iowa State University Press Series in the History of Technology and Science. Under the general editorship of Hamilton Cravens, the series operates in conjunction with the Center for Historical Studies of Technology and Science at Iowa State University. Overfield's volume follows Richard J. Blackwell's translations of Christiaan Huygens' *The Pendulum Clock,* Stephen G. Brush's *The History of Modern Science: A Guide to the Second Scientific Revolution, 1800–1950,* and F. Daniel Larkin's *John B. Jervis: An American Engineering Pioneer.*

I first learned of Charles Bessey some years ago when I read Overfield's exemplary essay in *Technology and Culture.* I later found myself at Iowa State University teaching in Bessey Hall. Overfield's essay prepared me to appreciate that coincidence. Overfield had taught me that Bessey was a person of considerable depth, a major figure within his own time.

Overfield extends that analysis here. He has not written his Bessey biography for the patently Whiggish reason that Bessey was an early proponent of Darwinian evolution or grasslands ecology. He did not choose to study Bessey because the scientist had connections to things we value or may be valued by some subsequent generations. Indeed, Overfield rejects any attempt to trivialize Bessey's life by reducing it to an explication of present concerns or by treating it as a milestone on the path to the present and future. Rather than fall prey to the temptation to make Bessey's life less than it was, Overfield has taken a consciously historical approach and concentrated on revealing Bessey's significance within his own era.

And what a glorious, important life that was. Bessey reigned as a preeminent agricultural scientist, botanist, and forester. He was hailed for his college teaching, university administration, and founding of the "new botany." His labors helped shape the Hatch Agricultural Experiment Station Act and the mission of the nascent land-grant colleges. He worked to expand the U.S. Department of Agriculture's scientific endeavors and was instrumental in the development of numerous professional organizations, including the American Association for the Advancement for Science and the Society for the Promotion of Agricultural Science. Bessey railed against machine politics and for free trade and public school reform. No wonder his contemporaries treated Bessey with admiration and deference.

Overfield's careful analysis of Bessey's multifaceted life helps us in another way. It enables us to begin to understand the time in which Bessey lived and flourished. It elucidates the distant time of the late nineteenth and early twentieth century. That is no small matter. Then-contemporary assumptions about nature and the nature of relationships among things have been obscured by the fog of time. Overfield's fine biography assists in that very crucial reclamation project. It is an impressive effort.

ALAN I MARCUS
Director, Center for Historical Studies of Technology and Science, and Professor of History, Iowa State University

# PREFACE

CHARLES E. BESSEY was an important person at an important time in the development of American science. His adult life fits neatly between the Civil War and World War I, a transitional period in which American society became "modern." As Robert Bruce has shown, science in America was established by the 1870s, but the next fifty years witnessed a maturing in many aspects. Evolutionary theory led a variety of new ideas and approaches that stimulated a reevaluation of the natural sciences and affected not only what one should study but also how one should investigate. The laboratory came to supplement the field and the textbook, and morphology and physiology were heralded as the avenues in which to revamp the natural sciences. The "new scientists" proclaimed the need to replace what they perceived as the old and inadequate practices and institutions, and the scientists of Bessey's generation built what they believed was a better foundation for science, its diffusion, and its applications. By the early twentieth century the structure that we identify as modern was in place and recognizable even though to later generations of scientists it too would appear inadequate and immature.

Into this general milieu, Bessey made his niche in a variety of ways. He shared the desires and ambitions of other scientists to make American science respectable in the eyes of Europeans and to convince Americans that science, if properly developed, would be a pillar upon which to build a modern and progressive society. In particular, he wanted botany to catch up with the other sciences. Botanically, Bessey's interests covered a broad spectrum including efforts to shift botany away from the traditional concentration on plant collection and description toward evolutionary classification, structural physiology, pathology, and ecology. This shift in emphasis was duly labeled the "new botany." Also, the "new botany" provided a foundation for professionalization and included improving education, scholarly journals, and organizations and societies. Bessey was a dominant figure in all these endeavors.

Equally significant, Bessey spent his entire career at land-grant universities and helped shape the nature of these fledgling institutions through nearly their first half-century of existence. This involved such issues as the relationship of the colleges to agriculturists, the beginning and development of agricultural experiment stations, and the struggles to provide a scientific basis for agriculture and agricultural education. Much of the focus of this study is on institutions, from local agricultural and horticultural societies and academies of science to science within the university. Nationally, Bessey aided the emergence of the U.S. Department of Agriculture as a major proponent of the "new science" and was active in numerous scientific organizations, the most important to Bessey being the American Association for the Advancement of Science.

As was typical of many early land-grant professors, Bessey enjoyed a broad range of interests and activities, which necessarily limited his contributions to each, although his "workaholic" habits still enabled him to be a productive scientist. At the same time, the wide range of achievements in science, agriculture, and education at such a key time in the formation of modern American institutions is a major justification for his biography. The coverage of each of these areas of interest concentrates on the issues and events that involved Bessey. Also, the sense of intimacy obtained from reading hundreds of his letters means that the description of the events of this period often reflects the viewpoint of Charles Bessey. Therefore, the perspective of this study tends to be that of a botanist at a land-grant college in the American Midwest. Such a perspective adds an important dimension to our understanding of American science where studies of the natural sciences have largely ignored botany, where agricultural and other applied sciences were not always considered science until recently, and where what was too often judged as significant in American science was restricted to a few eastern individuals and institutions.

The chapters are organized topically within a loose chronological framework. The first three chapters consider the foundations of Bessey's career as a botanist, the foundations of the land-grant colleges of Iowa and Nebraska, the development of agricultural science, and the foundations of state and local societies that had ties to the maturing of American science. These three chapters concern the 1870s and 1880s but illustrate the types of activity that occupied Bessey through the remainder of his career. Chapters 4 through 8 are largely topical and cover the major activities and contributions of Bessey's career—the "new botany," professionalization, understanding the prairies and Great Plains, agricultural botany, forestry and conservation, political and social reform, education, and evolutionary classification. These activities concentrate in the period of Bessey's greatest influence as a scientist on the national scene from the 1880s to about 1905.

Among the persons who have contributed to this study, I thank first David Sutherland and Roger Sharpe of the Department of Biology at the University of Nebraska at Omaha (UNO), who introduced me to Charles Bessey, and particularly to David for many helpful suggestions and corrections. Bessey's story is enhanced because he was an extremely active letter writer. Aside from part of his early career in the 1870s, he kept copies of correspondence received as well as written. It is this excellent collection of papers at the archives of the University of Nebraska–Lincoln, under the organization and care of Joseph G. Svoboda, that forms the core of this study and that breathes life into the events. The great accessibility to this collection and Joe's friendliness and cooperation greatly aided my work. A special appreciation goes to Alan I Marcus of Iowa State University for his encouragement and suggestions. Colleagues at UNO certainly deserve accolades for maintaining an enjoyable as well as stimulating environment and especially for enduring endless Bessey stories and occasional ventures into "hog history." Additionally there are many librarians and archivists who provided valuable help at the National Archives, Manuscript Division of the Library of Congress, American Philosophical Society Library, Missouri Botanical Garden, Department of Manuscripts and University Archives at Cornell University Library, Gray Herbarium, Library of The New York Botanical Garden, Linda Hall Library, Thompson Library of the University of Nebraska–Lincoln, and interlibrary loan office at UNO. Also, I thank Dean Margaret P. Gessaman of the Graduate College, the University Committee on Research, and the Department of History at UNO for financial support. Earlier versions of parts of this study were presented as papers to the History of Science Society, Missouri Valley History Conference, Northern Great Plains History Conference, and Midwest Junto for the History of Science. Finally, I acknowledge the University of Chicago Press for its permission to use material from my article "Charles E. Bessey: The Impact of the 'New' Botany on American Agriculture, 1880–1910," *Technology and Culture* 16, no. 2 (April 1975): 162–81, and the Forest History Society for its permission to use material from my article "Trees for the Great Plains: Charles E. Bessey and Forestry," *Journal of Forest History* 23 (January 1979): 18–31.

# SCIENCE WITH PRACTICE

# 1
# Laying the Foundations: The Iowa Years

CHARLES EDWIN BESSEY arrived in the small prairie village of Ames, Iowa, in February 1870 to begin a career in collegiate teaching that was to last for forty-five years. While he spent only the first fourteen of these years on the faculty of Iowa State College of Agriculture and Mechanic Arts (Iowa Agricultural College), they were the foundation years for Bessey as well as for the college. Building on his experiences in Iowa, Bessey was to make profound contributions to American science and agriculture during his long career as a researcher, a teacher, and an administrator.

Bessey already had a good preparation for the work that lay ahead when he arrived in Ames. Born on May 21, 1845, in Wayne County in rural north-central Ohio, he experienced the practical training of farm life. His father, a farmer during Charles's youth, had earlier received a classical education and had been a teacher. Bessey's parents stressed the importance of study connected with the classical curriculum, particularly of foreign languages, yet they believed that knowledge was important only so long as it was useful. The young Ohioan attended country school and at the age of seventeen ventured from the farm to attend district school and two academies from 1862 through 1866. Having acquired a teaching certificate during this time, he taught during vacations. In 1866, at the age of twenty-one, Bessey entered Michigan Agricultural College in order to become a civil engineer and surveyor. The Michigan college not only provided Bessey a good training in science for the time but also shaped his concept of what a land-grant college should be.[1]

The 1850s–1870s were a time of profound change in American higher education. A number of forces were at work, but two of the most prominent were the expansion and increasing prestige of science and the concept of the land-grant college. In many respects the two were closely connected. To some Americans, science was becoming more than the pursuit of interesting curiosities and hobbies and was providing answers to profound philosophical questions regarding the nature of the universe as well as to practical problems. Proponents claimed that science was the best—if not the only—path to knowledge and truth, and scientists claimed that an increasing number of problems lay within their realm of expertise.

With the actual and potential growth of science came demands for American colleges to provide more instruction in the sciences, and by midcentury college curricula included a varying amount of science, generally from one-fourth to one-third of classes offered. Still, educators made little effort to train scientists, stress research, or offer graduate study. In the few cases in which there were provisions for training scientists, college leaders did not place the courses, degrees, and students on the same level as the classical course of study and organized the science programs in separate institutions, such as the Sheffield Scientific School at Yale and the Lawrence Scientific School at Harvard.

While many of the educational reformers of this period agreed that there should be more science, they did not agree on what it should be. Should there merely be more courses in the basic training of an educated person; that is, replacing some of the classical courses in the program for the bachelor of arts, or should they direct the increased teaching of science toward vocational training? Mining interests were urging specialized training by midcentury, but only faint requests came from manufacturers and agriculturalists. Another scientific interest occasionally calling for educational change by midcentury was the desire to train scientists to engage in research. Combined, these three interests stated that a greater knowledge of science was necessary before a person could be considered an educated scholar, that the country needed more technology, that science would provide the best basis for technological advance, and that America needed persons with the ability and training to contribute to the scientific pool of knowledge. Thus by the mid-nineteenth century, however varied their intent might be, many proponents of a new education looked to science as their vehicle of change.[2]

The land-grant movement likewise had varied origins and intentions. Embodied in the movement were desires to break with classical education and provide what reformers considered a more up-to-date education; to remove higher education from the grasp of the upper classes and make it more than training professionals in law, divinity, teaching, and medicine; to make technology a legitimate college subject; to provide vocational training in a regular school rather than an apprenticeship setting; to introduce the German concepts of job training for technicians and civil servants and of research; to bring the government into open support of engineering and agricultural education; and to introduce new methods of teaching such as field work, demonstrations, and practical applications.[3]

The institutionalization of the land-grant movement came mainly with the Morrill Act of 1862. To a limited degree the movement was initiated by private groups and state governments that chartered agricultural schools in at least seven states during the 1850s. Few of these, however, were operational or beyond the initial stage of development before the Morrill Act. Some of the pressure in the states for more governmental attention to agriculture resulted in the formation in 1852 of the United States Agricultural Society to promote legislation. At the same time, a plan for industrial colleges in each state, formulated principally by Jonathan B. Turner of Illinois, reached Congress in the form of petitions from the legislature of Illinois and from an Illinois Farmers' Convention. These petitions called on Congress to provide public lands to support industrial colleges. Other proposals at this time called on the government to establish a national department of agriculture and a national experiment station. Congress, however, took no action on these proposals until 1857.

In December 1857, Representative Justin S. Morrill of Vermont introduced into the House a land-grant bill that would provide the following in each state:

> at least one college where the leading object shall be, without excluding other scientific or classical studies, to teach such branches of learning as are related to agriculture and the mechanic arts, in such manner as the legislatures of the States and Territories may respectively prescribe, in order to promote the liberal and practical education of the industrial classes in the several pursuits and professions in life.[4]

By 1859, the bill narrowly passed both houses of Congress but received a veto from President James Buchanan. With the withdrawal of Southerners from Congress and the election of a new president, a second land-grant bill received wide margins of approval in both houses and became law in July 1862. Most of the debate on the bill centered around the constitutionality of national aid to agriculture, the procedures of the land grants, and the effect of the grants on the states.[5] Little of the discussion concerned the educational aspects of the proposed colleges, but immediately with the initial implementation of the law questions arose as to what specifically a land-grant college should teach and how it should be organized.[6]

Michigan Agricultural College opened in 1857, nine years before Bessey started his studies there. The Michigan approach to applied science had a strong tradition within the state, particularly in the State Agricultural Society, and this approach was largely put into operation by the college's second president, Theophilus C. Abbot. The curriculum established the idea that a proper course of study for a land-grant college should consist of about one-third attention to classical studies and two-thirds to science and its application. Training in practical science should come only after a student was given a thorough introduction into the basic sciences—astronomy, geology, physics, botany, chemistry, and zoology. Also, the leaders at the Michigan school believed the goal of an agricultural college was to train leaders more than perfunctory practitioners.

By Bessey's student days, therefore, the course of study, which included no electives, combined the classical course with a heavy dose of the basic sciences. Abbot believed applications in science were built on experimentation and were based on a general knowledge of the basic sciences. Also at this time, agriculture was greatly emphasized over mechanics in the course of study. In mechanics, Bessey had applied classes only in surveying, industrial drawing, and civil engineering, while in agriculture he had semester courses in entomology, principles of breeding, horticulture, agricultural chemistry, and landscape gardening. In addition, there were chapel lectures each week on military studies and on such topics as horticulture, applications of chemistry, manual operations on the farm, and care and feeding of domestic animals. Of course, a student received much practical training through the manual labor program that Michigan Agricultural employed. Thus, under Abbot, although still in a financially precarious condition and still small in faculty and facilities, Michigan Agricultural College during the 1860s pioneered a middle path between classical and practical education, but a path definitely committed to agricultural education and science.[7]

The academic program at Michigan suited Bessey's scientific interests, although his specific attention as a student switched from engineering to botany. His training in botany at Michigan Agricultural was limited, but it was good for the times. In the reorganization of the college in 1861 when Abbot became president, the Board of Trustees hired Manly Miles, a zoologist, and George Thurber, a botanist. Thurber had a good knowledge of botany and an interest in agricultural botany. He left the college in 1863 to become editor of the prominent farm journal *American Agriculturist*. One of his students, Albert N. Prentiss, class of 1862, replaced him.[8] Under Prentiss, Bessey had semester courses in systematic botany and in structural botany and vegetable physiology, and in addition to the course in horticulture, Bessey had some practical training in Prentiss's greenhouse and model garden.

Receiving his bachelor of science degree in November 1869 at the age of twenty-four, Bessey stayed at Lansing during the winter term as an assistant in horticulture in charge of the college greenhouse. In December 1869, Iowa State College of Agriculture and Mechanic Arts (Iowa Agricultural College) hired Bessey as an instructor in botany and practical horticulture for the coming spring term for the salary of $1,000 plus a room in the student dormitory, a sum that was soon raised to $1,250.[9]

At Ames, Bessey encountered an educational institution in an even more youthful stage than he had experienced in Michigan as a student. Iowa Agricultural College was only in its second year of operation when he joined the faculty. The founders had rejected attaching the land grant to the already established state university in Iowa City and rejected placing the new college in the capital city of Des Moines.[10] Instead they had chosen, as Bessey later described it, a location "in a thinly settled part of the State, away from the railroad, and separated from a miserable little village by the almost impassable 'bottoms' of an uncontrollable prairie stream."[11] The isolation and the barren, treeless prairies most impressed Bessey on his arrival. In its eighteen months of existence, the college had come to consist of one large building, a farmhouse and sheds, and two faculty houses. As Bessey observed, there were "no drives, no walks, no paths, no smooth lawn, and only a few small trees."[12] One of Bessey's first tasks was to plan and carry out the landscaping of the new campus and to start a campus garden in an attempt to conquer the prairie.[13]

The faculty at the time consisted of nine men, including the president, Adonijah Strong Welch. Bessey lived for a short time in the Welch house, one of the two faculty houses located on campus. Welch had a diverse educational background, a background that blended the two traditions of classical and practical. He also had experience in and out of education and an executive ability that warranted the admiration of the younger Bessey. Therefore, from Bessey's standpoint, Welch was ideally suited to build a new college, and Bessey became one of the president's staunchest supporters in the attempt to pacify the varied interests in Iowa who disagreed on how the new college should operate.[14]

And there was considerable disagreement, both within and outside the campus community, on the type of education that was appropriate for a land-grant college. What constituted practical education as called for in the Morrill Act? To some persons the new land-grant colleges retained too much of the classical approach; to others, teaching how to plow a field or care for cattle was not legitimate college education.

In Iowa most of the criticism leveled against the agricultural college came from practitioners rather than from classicists; therefore, Bessey and others who supported Welch directed most of their energies to the various agricultural factions, trying to convince them that the college was proceeding in the correct direction. As Bessey pointed out, there were few models or textbooks or even formulated ideas of how higher education could best advance applied studies such as agriculture. Welch had investigated Michigan Agricultural College and based much of his program on that model. In Bessey, who provided Iowa Agricultural College with its motto, "Science with Practice," Welch found a person committed to the blending of three traditions—scientific, utilitarian, and cultural or classical. Because this particular view of practical education placed heavy emphasis on science, four faculty members in 1870 were in the basic sciences, three in applied areas, and two in cultural studies. Of course, the four in the basic sciences, like Bessey, devoted much of their time to applied teaching and some research.[15]

The Welch group, despite criticism that its members did not stress enough practical studies, was not trying to conceal a modified classical curriculum with merely talk and gestures toward useful knowledge. In fact, before 1877 the curriculum leaned much more heavily toward the practical than what the Welch group believed desirable. More than most other agricultural colleges, especially those attached to already existing schools, Welch led Iowa to break with the classical curriculum. Yet the Welch group firmly resisted those programs that Bessey termed the "cheap 'quick meal' type" of practical education, those studies pushed by some legislators and farm groups and in general by opponents of "book learning."[16]

To Bessey and others of the Welch group, the key to advancing agricultural education was science. In conjunction with other educational reformers, Bessey published his views on the need for a new agriculture based on the natural and physical sciences. Bessey believed the key to advancing agriculture was not to teach the student how to farm but by applying science, and because the sciences underlie any applications, these subjects were basic and had to come first for the student. In the long run, he concluded, agriculture would benefit more by "making the Arboriculture, the Botanical and Experimental Gardens, rather adjuncts to the natural sciences, than the reverse."[17] Bessey also recognized that college faculties were better equipped to teach the basic sciences than they were farming procedures. At the time, agricultural subjects lacked the organization necessary to present them successfully in the college setting, and Bessey believed that these subjects could be taught better on private farms than on any model farm or garden that the college, with its limited money and facilities, could develop. Therefore, according to Bessey, the agricultural college should concentrate on the basic sciences, award the student a college degree, and then provide an apprenticeship on a successful farm.[18]

Aside from arguments favoring practical training in farming, a major obstacle faced at this time by the Welch group was that proponents of the scientific viewpoint had to accept on faith that they had something important to offer in applied science. Researchers as yet had actually done very little to show they could help farming. The Welch supporters dedicated themselves to proving that their faith in science and in their concept of an agricultural college was justified. In this they were to be partially successful, but the process of making agriculture scientific was a long and groping adventure that did not show significant success until the 1890s.

Although initially uncertain as to how science could specifically benefit agriculturists, the Welch group moved to convince Iowans that the new college could be useful. One of the first attempts occurred during Bessey's first winter in Ames. President Welch initiated in four Iowa towns a series of meetings that became known as Farmers' Institutes. The intent was to have faculty members meet with and instruct agriculturists on varied practical farm subjects. Bessey participated in his first institute at Nevada, Iowa, in January 1871, the second winter that they were held. Immediately Bessey became a firm supporter of the idea and devoted considerable time to such meetings in Iowa and Nebraska during the next thirty years. Bessey regarded the institutes as one of the significant contributions of the land-grant educational movement, and relentlessly he attempted to quash claims by other colleges for the honor of pioneering such meetings. Later in his life he recalled, "Certainly we began this work in Iowa before they undertook it in any other state."[19]

The young botanist also was active in other ways in helping to bring the college to the people of Iowa in the early 1870s. He wrote short, informative articles on agriculture for Iowa newspapers and for farm journals, and he devoted much of his research energies to applied problems. Most of his work in the 1870s, however, was of a very elementary nature and consisted largely of collecting information or observing the results of simple experiments. Bessey developed exchanges with botanists from other regions of the United States and obtained from them such plants as potatoes and various types of grasses to try in the Iowa soils and climate. Bessey corresponded with outstanding agriculturists around Iowa endeavoring to learn their techniques of cultivation, their use of manures, the types of plants and seeds they used, and anything else their experience had shown to work. Bessey then organized and analyzed some of this material in order to offer it to farmers who made inquiries to the college. These were paltry attempts to justify the college and scientific agriculture, but Bessey was prevented from doing justice to experimental botany by his heavy teaching loads and by the lack of research facilities, funds, and a clear sense of direction.[20]

Subject matter in classes continued to perplex the members of the Welch group. Despite their general philosophy, they were not certain how best to carry out these ideas in the classroom and in the manual labor program. As with most of the new land-grant programs, there was difficulty finding persons trained in agriculture or in applied science. From Isaac Roberts and Millikan Stalker, who were largely responsible for the agricultural courses and the college farm at Ames, students received instruction in such subjects as preparation of soils, management of crops, and stock breeding. In related areas, Bessey offered classes in horticulture until November 1873 when he refused to include this and pomology in his chair of botany and zoology. Initially, all these courses were basically of the "how to" type with little attention given to the scientific underpinnings.[21] In 1874, the administration made another effort to satisfy the farm practitioners by establishing a one-year course in agriculture "consisting of farm specialties only."[22]

Despite these varied activities by Bessey and the other faculty, there remained criticism that the new college was not providing practical education or services. Critics levied charges against the Welch administration nearly from its beginning, but the discontent erupted openly and in a major way in 1873 when Lucian Hoggatt, a state legislator, and William D. Wilson, a Grange leader, asked for the removal of Welch and a reorientation of the college. These requests received an added boost with the exposure of the embezzlement of college funds and the firing by the trustees of three of the Ames faculty who opposed Welch.[23]

The three fired professors had been actively generating opposition to the views of the Welch group through writings in newspapers and through speeches and correspondence to some of the farm organizations within the state. In retaliation for their firings, the deposed faculty intensified their charges against Welch and the college and provided a rallying point for farm opponents of the college, particularly the Iowa Grange. The result was the formation of a special legislative committee charged with investigating the accusation that the college was not "fulfilling the purpose for which it was founded." Regarding the general question of the college's intent, the anti-Welch faction charged that the faculty and the curriculum encouraged students to train for nonfarming occupations, that the methods of instruction were not oriented toward practical learning, and that the college farm was not a model farm but used only for research. During the investigation, which lasted more than a month, the committee interviewed forty-three college and state officials, faculty, students, and interested citizens. The charges were serious, and the committee delved deeply into the workings of the college from its opening onward. The charges were a threat to Welch personally and to the emerging concepts of scientific agriculture in general.[24]

When the fracas broke, Bessey was in Massachusetts for his wedding, and Welch wrote the botanist of his concern over the investigation. "It would be idle," Welch warned, "that our side is trusting too much in the merits of their case." Welch reported that the committee was organized along politically partisan lines with four "strong men," all Republicans, and four Anti-Monopolists or Grangers. The seriousness of the affair was evident to Welch, and in closing he pleaded with Bessey as one of his most able and loyal colleagues, "You should come back at once; come here yourself and help me."[25]

Bessey returned to Iowa and testified near the end of the investigation. Most of his statements and the questions directed to him involved his work with the college garden and with the practical nature of his instruction. As expected, Bessey defended the teaching and the manual labor system as used at Ames. He emphasized that the work in the garden was useful to students and that he had spent three to four hours daily supervising students until the present year's heavy teaching schedule kept him from continuing. Likewise, Bessey believed that instruction in entomology, plant physiology, comparative anatomy, agricultural chemistry, and drainage sufficiently stressed utility.

When asked to compare his experiences in Ames with those of his student days at Michigan Agricultural College, Bessey acknowledged that manual labor at Lansing was a little better organized because Michigan Agricultural was an older institution and the faculty had had longer to develop a meaningful work program. On the whole, however, Bessey considered Iowa Agricultural superior to the Michigan college in having more instruction and greater diversity in its manual labor. As for agricultural instruction, he believed it was improving at Ames, and he rated Iowa graduates better overall than those from Michigan because of the improved curriculum and a higher-caliber faculty.[26]

The findings of the committee supported Welch against the personal charges, but there was enough indecision by the committee regarding the general questions that the controversy over the nature of the college did not end. The committee found no fault with the administration's handling of finances other than the embezzlement by the treasurer, which was really not part of the charges against Welch or the college. Likewise, it found no evidence of mistreatment of students. The committee agreed that the board of trustees had acted legally in dismissing the three faculty members, but it divided over the "manner, necessity, and propriety" of the firings. On the general question of whether the college was corrupting the intent of the Morrill Act, the committee had heard a good deal of confusing and conflicting testimony, yet it concluded that the curriculum and the instruction were in line with the original purposes of agricultural education. Regarding specifics, however, the committee was less conclusive. For example, the committee found the work of the model farm oriented toward experimentation rather than demonstration, but it could not agree whether that was good or bad.[27]

Although demands for the time of the early land-grant professor were great, Charles Bessey had the ambition and energy to lay the foundations for another aspect of his career, that of a scientist. Even though Bessey spent his entire life close to agriculture and contributed to its academic and scientific growth, he never considered himself an agriculturist. He regarded himself always as a scientist, more specifically a botanist.

Bessey was enthusiastic about the growing efforts to bring science into greater prominence in American life, and as did many other young professors in the 1870s, he became a missionary for the movement. One of Bessey's first and best friends at Ames was W. H. Wynn, a professor of literature. Later in his life, Wynn described the mood of that early period: "In those days science was in the very dawn of an epoch which was destined to become all its own. The excitement about evolution and cognate mysteries was worldwide and wild." Wynn placed Bessey in "that great galaxy of scientific men" who were revolutionizing thinking, and he characterized the botanist as "active of mind . . . demonstrative in class and out . . . awake to everything going on," and a person who never rested—"he lectured on botany everywhere and anytime."[28]

Science was even brought into the social life of Ames. A group of townspeople including Bessey, Wynn, and Welch formed a social club named the Lamellibranchophagi, or Oyster-Eaters, whose members met every two weeks, ate oysters "served in every style," and listened to short lectures. At the gatherings, according to Wynn, Bessey was the catalyst and the "air was literally thick with science, and the feasting of the society was ordered in heavy scientific turns."[29]

Wanting to become a good botanist, Bessey looked for guidance and—following the pattern of many other young Americans interested in botany—he turned to Asa Gray, the "patriarch" of American botany and professor at Harvard.[30] At this time, Gray was one of two full-time professors of botany in the United States, and he headed the only facility in the country equipped and staffed to offer graduate training in botany. In answering Bessey's occasional letters, Gray advised the young botanist on a variety of topics. He recommended the best botany textbook for the Midwest, answered technical questions about plant sexuality, listed some eastern botanists who would exchange plants, suggested the *American Naturalist* as the best journal to read and to which to submit his research results, and encouraged Bessey to send Iowa plants, particularly wild specimens, to Harvard.[31]

Bessey followed Gray's advice, including suggestions on topics of research, and during the next several years the Iowan had his first scientific efforts published in a national journal, though admittedly they were only short notes describing simple observations of plant locations and ranges of growth.[32] A main item of interest in their early correspondence was Bessey's discussion of the peculiarities of the compass plant, so called because its leaves lined up in a north-south direction. Gray encouraged Bessey to pursue an explanation for this characteristic, and in a little over a year Bessey had developed a procedure for studying the polarity of the plant, which Gray volunteered to submit as a paper at the annual meeting of the American Association for the Advancement of Science (AAAS). Bessey worked several more years before he determined the nature of the leaf structure that oriented the leaves in its peculiar fashion to allow both sides to have equal exposure to sunlight.[33]

In the meantime, in August 1872 Bessey had his opportunity to meet Gray in person at the annual AAAS meeting in Dubuque, Iowa. Gray came to deliver the presidential address, and Bessey attended, was elected to membership, and despite the poor attendance made some acquaintances in the American scientific community.[34] Bessey did not record the nature of his conversation with Gray, but they did outline the arrangements for Bessey to spend his three-month winter vacation studying at the Harvard College Botanical Garden. Having just received an honorary master of science degree from Michigan Agricultural College and been promoted from instructor to professor, Bessey was eager to advance his career and Harvard was the right place to do so. Gray was in the midst of restructuring the botanical setup at Harvard to provide better means for teaching and particularly to provide instruction in botanical areas beyond taxonomy.[35] Gray set Bessey to work organizing and classifying a set of plants from California, and in addition, Bessey was exposed to the teaching of George L. Goodale, a new member of Gray's staff and an advanced student in the developing field of plant physiology. Bessey also worked in the new laboratory for fungi, even though William G. Farlow who handled this instruction was in Europe.[36] Not only was Bessey's basic knowledge of taxonomy improved, but also his experiences with physiology and the fungi greatly broadened his understanding of the larger scope of botany.

Bessey later recalled his studies with Gray. Then in his early sixties, Gray had assigned Bessey to work largely on his own in the laboratory, making his way through key groups of the plant kingdom. Bessey always was impressed with the cheerfulness and good humor of his mentor and found Gray's infrequent lectures "inspiring, carefully prepared, given offhand or from outlines, simple even in technical subjects, full of enthusiasm, never dull, never slow and halting." Yet despite Bessey's admiration, he noted that the "boys of that day," the undergraduates, "did not appreciate him [Gray], if he was late they would cut the class."[37]

While conducting his studies in the East, Bessey met Lucy Ahearn of Martha's Vineyard, Massachusetts, and they were married on December 25, 1873. Professionally Bessey made a good enough impression that two years later Gray suggested him for a temporary teaching position at the new University of California. Daniel C. Gilman, president of the university, asked Bessey to write about his ideas on the need for agricultural science and education.[38] From their correspondence developed an invitation for Bessey to give a course of lectures on botany in California during his winter vacation in early 1875. This was an excellent opportunity for Bessey to look firsthand at the flora to which Gray had introduced him in the Harvard herbarium two years earlier.[39]

In August 1875 after his return from California, Gray inquired as to whether Bessey was interested in leaving Iowa because persons at Cornell and California had asked about him. As the pressures on the college in Ames continued and since Bessey was making good progress toward establishing himself as a promising botanist, he became more interested in leaving Ames. Bessey summarized his feelings by stating that he hoped a "real college" could be fashioned in Iowa, but if this appeared impossible he wanted "to avail himself of the first opportunity to secure an appointment in a college with a recognized position or with a hope for the future."[40] California particularly interested him. It was a new school in which he hoped the "old ways" would not suppress the "new" education, and botanically the area was rich and largely unworked. In early 1876 Gray passed on the rumor that California was prepared to establish a chair in botany, and Bessey applied for the position only to find that financial problems prevented any additional faculty appointments.[41]

Although apparently disappointed over the situation in California, Bessey, now thirty years old, continued to study botany. During his winter vacation in early 1876 he returned to Harvard. Again, his primary

## LAYING THE FOUNDATIONS: THE IOWA YEARS

studies aside from the standard taxonomy were in structural physiology and the nonflowering plants.[42] One of his first physiological investigations was to observe the effects of cold temperatures on protoplasm in plant cells. He had made some observations since first coming to Iowa, but it was during this second period at Harvard that he used the laboratory facilities to repeat a number of the tests described in the European literature and to look at some additional plants on his own. Bessey's work in physiology at this time was primarily of the type that required the use of the microscope to study cellular and tissue structures and to explain the functions of these structures.[43]

Similar investigations that Bessey pursued in the mid-1870s involved the effect of heat and light on color in flowers and attempts to measure growth in plants.[44] He built some instruments and improved on those used in Europe that he could put to use in a physiological laboratory.[45] Another area of research that greatly interested Bessey in these early years was reproduction in plants. He had for several years studied touch sensitivity of stamens to find how various plants bent their stamens upon contact with insects to insure covering the insect with pollen, and he investigated the heights and structural relationship of sexual parts to allow or prevent self-pollination. While these studies were small in scope, he was able to correct Gray and other botanists regarding the reproductive structure and function of several species.[46]

Although he had been interested in plant structure and functions since his undergraduate days, Bessey by the mid-1870s was also developing a strong curiosity about "lower plants," particularly fungi. Much of this work and interest again allowed him to apply his growing facility with the microscope to learn more about the general nature of fungi. Aside from his scientific curiosity, Bessey recognized the potential economic importance of such organisms to agriculture. At this time, botanists knew little about such crop-damaging organisms as smuts, rusts, and blights, and the little that they knew pertained to the northeastern United States. In most instances, knowledge of these lower groups was not even sufficient to provide the basis for classification. Almost nothing was understood about their physiological nature.

Bessey realized that until botanists knew more about the basic form and workings of these organisms, it was impossible to proceed with much applied work. The fundamental scientific knowledge necessary for practical applications was so little understood that there was no consensus on how, if at all, fungi affected other plants. There was uncertainty regarding terminology and even disagreement about whether rusts and smuts were terms for the same or for different organisms.[47] Spores and the life cycles of these harmful plants remained largely a mystery. Botanists still disagreed about whether rusts, smuts, and molds were the cause of plant diseases or merely side effects of diseases. Even greater disagreement existed regarding the question of bacteria's role in disease.

What further complicated the study of possible correlations between "lower plants" and diseases was the lack of agreement on basic questions about the nature of diseases. Was a disease the result of specific causes or of general causes such as climate, structure, physiology, or soil? Once a scientist accepted that fungi caused plant diseases, as Bessey did during the 1870s, a vast gap in knowledge remained to be bridged before the researcher could approach the problem that concerned the agriculturist—that is, how to prevent these diseases. Although Bessey believed that a specific fungus caused a specific disease in plants, during the 1870s he did not yet accept the idea that bacteria caused diseases.

In contrast, an Iowan who described himself as a practical farmer in response to an article by Bessey believed that diseases such as pear blight resulted not from specific organisms but from the plant being moved from its natural soil and climate, thus causing "an abnormal or defective growth." The farmer added that an excess of organic matter in the soil combined with "certain conditions of heat and moisture" to produce rusts, smuts, and blights that injured crops.[48] This was an example of the general theory of disease which, although quite popular during the previous half-century, was being challenged by the 1870s.

In light of these conditions, it was evident that when Bessey turned to problems in Iowa, he had in most instances to start from the beginnings, which to a botanist meant he had to systematize his material, to collect and classify.[49] Asa Gray and John Torrey's *Flora of North America,* the major floral description of the country available at the time, did not have the depth and range for more than an overview of Iowa and the Midwest. Since his first days at Iowa Agricultural College, Bessey had harbored the idea of a survey and catalog of Iowa plants as a necessary first step in classifying the type of vegetation that grew in the region. Immediately he started a college herbarium, began to collect plants, and—to encourage others within the state to follow his lead— published what he hoped was the first installment of a state plant survey. Next the botanist sought to obtain public support for his project. He failed, however, to have botany included in the existing State Geological Survey, and separate public funding seemed even less likely.[50]

In lieu of recognition and financial support for a botanical survey, Bessey encouraged interested persons within the state to collect and classify the plants of their locality. Bessey informally organized these persons, asking them to make plant lists and submit them to him, hoping that the information soon would appear in the biennial college reports or in publications of the state agricultural or horticultural societies. Certainly Bessey recognized the flaws of such an arrangement. The work of the persons involved was very uneven, and most of the results were no more than lists of plants by scientific name only. Nearly all these persons were amateurs—many could do little more than collect the plants—and the project required too much of Bessey's limited time to identify the specimens properly. The major weakness of the system was the sporadic nature of such a survey. There was no way systematically to cover the state and the many types of plants that existed, but it was a start and it helped awaken agriculturists to the importance of botanical studies.[51]

Using the college biennial reports as a means of persuading Iowans of the need for an expanded public plant survey, Bessey in 1873 noted that the college herbarium now contained over five hundred species and that the catalog of state plants was over one thousand and in need of a permanent means of publication.[52] Continuing to give special attention to the mosses, lichens, and fungi of Iowa, Bessey had organized a preliminary checklist of each of these groups by 1876. He stressed the incompleteness of the lists, however, by pointing out that the lichen catalog contained only about one-fifth of the estimated species within the state.[53]

The first published results of this work came in 1875 when Bessey included the article "On Injurious Fungi" in the college biennial report. He added a second installment in 1877.[54] As with the mosses and lichens, these two fungi studies were basically a listing of species, although he did add a short description for each. The descriptions were of a very uneven character, but as Bessey explained, there was so little known or agreed upon regarding American fungi that he had to rely primarily on the few sources available: the works of Harvard's William G. Farlow, England's M. C. Cooke, and New York's Charles H. Peck. Where he could, Bessey added his own notes to the descriptions, particularly for the common Iowa species. To make the list more useful for farmers, Bessey added comments about how each species harmed its host and what plant served as host. Despite his efforts, Bessey was unable to advance beyond these lists and short descriptions during the 1870s.

While establishing himself as a botanist and working to provide a scientific foundation for education and agriculture, Bessey joined with other interested persons to create institutional homes for science within Iowa. Bessey envisioned colleges as the primary repository for science, but he recognized that the boundaries for scientific investigation must not be drawn too tightly because science inside the colleges was not greatly advanced beyond the general level of educated persons at that time. In particular, institutions outside the colleges were needed to promote the ideas of applied science and to provide the pressure groups that could successfully secure public financing. In the early 1870s Bessey's primary choice as a home for practical science in the state was the Iowa State Horticultural Society.

Representatives of the horticultural society were the most enthusiastic proponents of scientific agriculture within the state. Thus, plant physiology, meteorology, geology and soils, insects, and birds were the areas of science most frequently discussed by the members, and the society, which operated largely by committees, organized itself around these topics. The committees consisted of one to three persons who largely worked independently on reports for the annual meetings. These reports then constituted the papers presented at the meetings and published in the annual report. In their annual addresses, most presidents of the society discussed the need for more science in horticulture, and the secretary, J. L. Budd, discussed the subject each year in his report and in papers to the society.[55]

Members of the Welch group were active in the society. In his reports to the state organization, Bessey always stressed the role of science in the work of the society, and H. H. McAfee, who took over the duties as professor of horticulture from Bessey in 1874, provided numerous excellent statements on the potential of practical science during the several years he was at the college. President Welch also actively participated in the annual meetings, and although most of his contributions were to the committee on landscaping, he occasionally presented general addresses. All three men tried to demonstrate the necessity of science for the improvement of horticulture, and in general their pleas fell on a receptive audience.

Although the members of the State Horticultural Society were committed to science, they were extremely vocal that little was being done to advance knowledge beneficial to horticulture, and in expressing their displeasure the agricultural college took much of the criticism. Concerned with the lack of available scientific information, the horticultural society passed a resolution in 1871 calling for part of the college farm to be "especially adapted to experimental pomology and forestry."[56] As the president of the society stated two years later, experimental horticulture "properly belongs to our Agricultural College and from that we have a right to expect much more than we have ever realized in this line."[57]

In response to these feelings, the society established a special committee to inquire into the necessary cooperation between its organization and the college, and during January 1874 the committee confirmed the opinion that experimentation was a proper function of agricultural education. Criticism of the college for its lack of experimentation usually was balanced with a recognition that the college was new and was not given proper funds or staff to adequately carry on necessary and extensive testing.

From the college side, members of the Welch faction continued to address the society and to assure its members that the college recognized its obligations in this area. In 1875 in "The Relation of the Agricultural College to Horticulture," President Welch summarized his views on the general intent of the Morrill Act and how his faculty were meeting these intentions. The legislative investigation of the college had just been conducted the previous year and Welch was still under tremendous pressure to justify the actions of his administration. Outlining the primary purposes of the college as instruction in the agricultural and mechanical arts and teaching the children of the working class, he added that while the law allowed classical and general scientific studies as added features of the education, he was "more and more in favor of the strictest construction" of the law. He also believed the Morrill Act "evidently contemplates experimentation on agriculture whenever the leading objects are obtained." Welch explained that the requirements of developing the instructional aspects of the college had absorbed nearly all the money and energies available to this point, and he characterized the beginning years of the college not so much as development as "self-preservation." These conditions, claimed Welch, explained the inability of the college to provide horticulturists with the type of work the faculty knew was needed to make the college useful. In his general defense of the college, however, Welch fell back on his basic belief that a proper agricultural education would produce better farmers but more importantly would train persons in scientific agriculture. "Agriculture can be helped in ways other than actual engagement in raising of crops," he concluded. "This is a world where skill alone, can gather the rich ripe fruits of industry, and therefore the new education seeks to impart to the student the skill of the expert."[58]

Despite their favorable attitude toward science, members of the society concluded that little scientific information was available that would actually produce better plants. C. C. Parry, a member from Davenport, Iowa, and one of the country's leading plant collectors, emphasized this, stating that "from the storehouse of American botany, horticulture has hardly commenced to draw in a proper way the material so much needed to enrich her domain."[59] Professor McAfee succinctly outlined to the members a more accurate picture of the limits of scientific horticulture. "This imperfect condition of the science, Botany," he explained, "is exactly what stands in the way of the improvement of varieties which all Horticulturists desire. We desire to do, but don't exactly know how to do. Perhaps too much energy has been applied to classifying forms of natural growth, and not enough energy has been given to investigations into plant physiology." To McAfee the problem was the narrowness of botany with its attention on taxonomy. When the horticulturist goes to the botanist for information, queried McAfee, what does the botanist say? "Why he tells us that the Apple is *Pyrus Malus* . . . he endeavors to classify, and classify, rather than to tell us how *Pyrus Malus* of to-day may be made more delicious, more hardy, more productive." McAfee concluded that the answer for the horticulturist was the "new biology" of Charles Darwin, Herbert Spencer, John Tyndall, and Thomas Huxley, which in botany emphasized "how plants originate, grow, develop, and mature." Plant physiology thus was the part of botany that should interest horticulture.[60]

As Welch, Bessey, and McAfee hammered away in their defense of the college and their belief that science must precede practice, members of the State Horticultural Society were generally very understanding regarding the conditions at the college, and they became a significant pressure group to obtain more money, time for the faculty, and an environment that encouraged and rewarded experimentation. They realized that the lack of expertise in practical agriculture was not unique to Ames but was true throughout the country. In general the criticism by the horticulturists of the lack of research became not so much a criticism of the college but of the state legislature for not providing more money and better facilities. The State Horticultural Society also spoke out for the need to overcome the limitations of "random experimentation" and called for a systematic research program that could best be carried out by a state experiment station. The society believed that the farm of the agricultural college was the ideal location for this program. Starting in 1878 and repeating every year after for the next decade, the society went on record supporting the establishment of such a research facility.[61]

In line with the type of work Bessey encouraged in the State Horticultural Society, another of his major interests during the early 1870s was his attempt to obtain a state entomologist for Iowa. As a sideline to his main activities, Bessey wrote several papers on beetles and on harmful insects in Iowa.[62] He recognized, however, that the little time he could devote to the subject was woefully insufficient, and yet the topic was of immense practical importance to farmers. He wanted the college to add a department of economic entomology, and he encouraged the Iowa State Agricultural Society and the State Horticultural Society to pressure the legislature to establish the position. For as little as $3,000 annually, Bessey pleaded, the government could save Iowans many times that amount in crop losses to insects. Bessey believed that both national and state governments were proper agents to finance agricultural science because in his opinion, nothing could promote the general welfare more than public support of agriculture. C. V. Riley, state entomologist of Missouri and a member of the U.S. Entomological Commission, frequently corresponded with Bessey, advised him on the needs of insect studies, and was willing to move to Iowa. But Bessey found the legislature unmoved by the pleas for public support of agricultural science in the form of an entomologist. He considered his unsuccessful experiences in securing a state entomologist characteristic of the sad condition of public science in general in the 1870s, a condition which was only too evident in the U.S. Department of Agriculture (USDA) at this time.[63]

Much of Bessey's other involvement in the State Horticultural Society was of an indirect nature during the 1870s. The annual meetings were always in January, which coincided with the three-month winter vacation of the agricultural college, and during the four years from 1873 through 1876, Bessey was gone from Iowa at that time of year to study or to teach. Although absent, he wrote committee reports and had them read and included in the proceedings. His early committee work was quite varied, ranging from meteorology to experimental horticulture, but as expected, his primary attention was to the botany committee. He presented short papers on various aspects of plant physiology including diseases, the nature of protoplasm, and the effects of low temperatures on plants.[64] By the mid-1870s, the committees were becoming larger and narrower in subject, and Bessey was gone from the meteorology and experimental horticulture groups. He recognized their work as essential to the society but he was happy to relinquish his seat as soon as there were other qualified members. He was then free to concentrate on botany, particularly on plant diseases.[65]

Although his work in the horticultural society was important to the growth of agricultural science in Iowa, Bessey played an even greater role in trying to provide a home within which science in general could grow in the state. While he envisioned the agricultural and horticultural societies having an important scientific position, Bessey also wanted a home for the study of science for its own sake without the obligation of usefulness attached. Out of this desire, Bessey joined a small group of similarly minded scholars in 1875 to found a state scientific association, the Iowa Academy of Sciences. Earlier, Bessey had hoped for a local botanical society or a state natural history society, but he came to realize that the scientific community in Iowa was too small to allow narrowing the group into specialized areas of science. Therefore, the academy was open to the broad vistas of the natural and physical sciences, although it did exclude the social sciences.[66] In August 1875 seven persons met in the university laboratory in Iowa City under the leadership of Bessey. Those present selected five additional persons to constitute the charter membership. This charter group consisted of seven physicians and five professors, with Bessey serving as president.

Only five members attended the academy's first semiannual meeting in January 1876. Bessey, who was studying at Harvard, missed. The few present discussed a constitution, approved five new members, and listened to the reading of one paper. Bessey sent two papers for the meeting, "Some Observations on the Flora of Iowa" and "Contributions to the Bryology of Iowa," but only their titles were read. The meeting was not an illustrative start for the academy, but it was to be typical of meetings for the next decade, at least from the standpoint of attendance.

Bessey and five other members were present at the meeting in June 1876. They reelected Bessey president and approved the constitution. The constitution embodied the desire to limit the academy to a scientific elite. They restricted membership to those persons who had performed "good scientific work," which to Bessey meant research and the writing of scholarly papers. Members were dropped if they failed to attend or to present a paper every two years. Membership also was limited to thirty persons, a limit which the academy never reached during Bessey's residence in Iowa. Small attendance, papers often less than first-rate, and the lack of money to initiate the publication of their proceedings attested to the shortage of scientists in the state, yet they had established, although only temporarily, the foundations of a home for the promotion of science in Iowa.[67]

Thus by the mid-1870s Charles Bessey had helped formulate and implement the direction the Welch group desired for a practical, yet scientific, agricultural education at Ames, but it still faced substantial opposition. He had taken significant steps that he hoped would bring him into the mainstream of the expanding discipline of botany, yet he was barely known outside Iowa. And he encouraged Iowans to do more to provide the institutional basis for practical and pure science within their state, but little had been accomplished outside a small circle of agriculturists and little headway had been made with the state government. But it was a period of pioneering in Iowa, in science and education as well as in other facets of life, and Iowans were fortunate to have had the services of Bessey available to them at such a time.

# 2
# A Time of Promise but Uncertainty

BELIEVING that scientific agriculture held the best promise for the future and success of practical education, a small group of faculty and interested citizens provided the initial groundwork for Iowa Agricultural College during the early 1870s. Yet progress in science at the Iowa college was slow since agricultural education was in its infancy. Members of the group that supported President Welch also were slow to implement their ideas about scientific agriculture because there was strong opposition to them within the state. It was not until 1877, therefore, that the faculty at Iowa Agricultural College moved to bring their ideas into reality and began to gear the curriculum at Ames more toward the study of scientific agriculture.

The new plan required students to study the basic sciences with the idea that a thorough knowledge of science would allow students in advanced courses to apply specifically to agriculture the general scientific principles learned. The courses were intended to teach students why to use fertilizer and what materials to choose for best results, but not how to apply the fertilizer. Agricultural students now had to take the same science classes as did students preparing for the bachelor of arts degree, and by 1880 the agricultural program was elevated to the point where students in it were eligible for a bachelor of science degree. These changes fit better with the philosophy of the Welch group.[1]

As part of their curricular changes, the faculty also modified an important part of the traditional training in agricultural colleges: manual labor education. Aside from the assumption that manual labor was a healthy and character-building activity, such work in an agricultural college provided opportunity for practical training. At Ames, however, after the initial establishment of the campus farm and grounds, the faculty found few manual duties that they considered educationally beneficial.

As a result, the role of manual labor at Iowa declined and was modified until the college finally abandoned it altogether in 1884.² By 1876 the college had quit paying students for work classified as instructional, and the faculty gradually had replaced the educational side of manual labor with another type of on-the-job training, the laboratory method of instruction.

It was in the introduction and expansion of the laboratory method of instruction that Charles E. Bessey made one of his greatest contributions to the land-grant concept of education and to the teaching of botany in general. Before the 1870s the laboratory method of instruction in the United States was restricted largely to chemistry and physics and even in these fields was used only sparingly. A respect for the scientific schools of Germany by Bessey and other scientists of his generation helped expand interest in the laboratory.³

Some confusion remains about when Bessey first introduced his botanical laboratory. In February 1871, during his second year at Ames, he had his botany students collect, dissect, and mount specimens using a hand lens in their analysis of plants. Two years later Bessey moved to formalize his laboratory. He closed off the end of a hallway, provided a table and the one compound microscope that he had persuaded the administration to purchase, and placed the title "Botanical Laboratory" over the door. Because these were days of conscious American pioneering in the sciences, Bessey often has received credit for starting the first botanical laboratory in the United States. More accurately, Bessey claimed only the first botanical laboratory outside Harvard, but he did claim the first laboratory for undergraduates. However, a later student and colleague of Bessey's, Raymond Pool, stated that the use of the graduate laboratory at Harvard—one year prior to the establishment of Bessey's botanical laboratory—was unknown to Bessey at the time.⁴

Admittedly Bessey assigned very little laboratory work to his students during the first few years because he had to learn how to fit such activity into a traditional textbook subject and had to provide his own materials. By 1874, however, the laboratory became an established and basic part of his teaching.⁵ Bessey's pride in this venture was evident in a letter he wrote in 1915 to alumni from the Iowa class of 1875: ". . . before that there was but one botanical laboratory in America, at Harvard and that only established a year or less, and yet on the Iowa prairies in a college not half a dozen years old you were taking part in a biological revolution that has swept the country."⁶ Bessey was not exaggerating in this claim. By the 1890s the laboratory method of instruction had become the accepted way of teaching botany, and botanists throughout the country acknowledged the foundations provided by Bessey at Iowa Agricultural College.

Aside from the simple analysis with the hand lens, Bessey centered his laboratory work around the compound microscope. He remembered that the botany room at Michigan Agricultural College had contained one compound microscope, but only once, and on his own, had he used the instrument as a student, for it never was offered as part of the botanical instruction. By the early 1880s Bessey had acquired enough microscopes at Ames for the students to use them on a regular basis. He fondly recalled that the faculty and administration generally believed "that the professor of botany was slightly 'queered' or out of his head when the subject of microscopes was under discussion." President Welch, Bessey added, "never really fully understood my insatiable thirst for buying more microscopes."[7]

Bessey always stressed to Welch that money for the laboratory was much more important than money for the fledgling museum, which traditionally had been associated with teaching natural history and which had been one of the first fixtures at the new Iowa college. According to Bessey, the emphasis in botanical education needed to be shifted from the museum to the laboratory because in the museum a student could look only at dried or pickled specimens but in the laboratory could touch and work with living organisms.[8] Bessey early adopted a principle from Louis Agassiz, one of America's leading proponents of scientific development, that the key to learning biology was to "study nature, not books," and Bessey preached this motto to anyone who would listen.[9]

Thus, by the early 1880s Bessey's enthusiasm for the laboratory method was spreading through its use by others who had similar experiences and by his students who became teachers in both secondary schools and colleges. Bessey personally introduced the botany laboratory while teaching at the University of California in 1874–75 and the University of Minnesota in 1881.[10]

In 1878 Bessey moved into a new phase of his budding career when he turned to writing textbooks. He received a brief introduction to this activity when he was approached to revise the volume *Geography of Iowa*. This assignment was primarily a matter of money, however, and he quickly wrote the brief twelve pages of text and turned to botany.[11]

Henry Holt and Company had planned a new science series for high schools and colleges for four years, but by 1878 they had nothing published. George L. Goodale of Harvard College had agreed to write the botany text, but Holt was dissatisfied that he had not made the book "sufficiently general." Unable to settle their differences, Goodale withdrew and suggested that the company contact Bessey to replace him. Goodale believed that the series of lectures Bessey had written and delivered at the University of California in 1875 could be what Holt wanted for its textbook—that is, what "every educated gentleman should know."[12]

The editors of Holt informed Bessey that they intended the series for general education and not for the specialist. In updating their science textbooks they wanted an American orientation and original illustrations; in particular they wanted a textbook "in accordance with the doctrine of evolution." They emphasized, however, that due to "prejudices" that still existed they would not advertise the stress on evolution.[13] Bessey was to receive ten percent of the retail price for every copy as long as there were more than fifty copies sold within each year. Holt desired the manuscript completed in a few months, Bessey promised it in eight, but the book actually took two years to write.[14]

Although he had already agreed to write the textbook, Bessey felt obligated to ask his Harvard mentor Asa Gray if he should undertake the project. "I am painfully aware of my shortcomings in historical knowledge," Bessey admitted, but he concluded that "I suppose it would do me a great deal of good in the experience I will gain." Gray replied, "I wish you success but [I] must not do more for your Holt and Company book because it is intended to be one of my rivals in that field, which is why Goodale could not touch it."[15]

Revising his California lectures for the textbook forced Bessey to shift the nature of his own study. For the next two years he mainly reviewed the important botanical literature, and aside from an article on apple blight, the papers he prepared were basically material he was using in the book.[16] When his eight-month deadline arrived in January 1879 he asked for an extension, and delays continued to mark the writing. Thus, it was December 1879 before Bessey finished the manuscript, and it was August 1880 when the book finally was published and ready for marketing. The end product was much more than Holt had intended.[17]

When finished, Bessey believed that he had created a distinctive textbook. In addition to serving as a traditional introduction to plants, the book was designed to help the student "become an observer and investigator and obtain knowledge first hand." As far as interpretation was concerned, he claimed that the only "considerable innovations" he included were two: (1) recognizing seven distinct types of tissue, and (2) recognizing four groups of "lower plants" (Protophyta, Zygosporeae, Oosporeae, and Carosporeae) as plant divisions on the same level as the three generally recognized major groups (Bryophyta, Pteridophyta, and Phanerogamia).[18]

Bessey's *Botany for High Schools and Colleges* entered a market dominated mainly by Gray, Alphonso Wood, and Almira Lincoln Phelps. Typical of American botany books of the time, these texts used British studies as their model.[19] Although drawing much of his information from the English botanist William Ramsey McNab and basing his classification system largely on the work of Britons George Bentham and Joseph Hooker, Bessey used German studies more for his outline of how to study botany. The chapters in Part I on anatomy and physiology in particular, Bessey acknowledged, followed "nearly" the study *Lehrbuch der Botanik* published by the German botanist Julius von Sachs in 1868 and 1874.[20] Bessey had been eagerly studying and absorbing German botanical writings ever since Harvard's Farlow and Goodale had most likely introduced him to them. Bessey acknowledged his reliance upon scholarship by German and Swiss physiologists such as Wilhelm Hofmeister, Anton de Bary, Carl Wilhelm von Nageli, Eduard Strasburger, and Simon Schwendener. Certainly one of the main strengths and unique features of Bessey's textbook was its use of current European research.[21]

Of Bessey's chief competitors, Asa Gray, in his *First Lessons in Botany and Vegetable Physiology* in 1857 and his *Introduction to Structural and Systematic Botany* in 1858 (a revision of his 1850 *Botanical Textbook*), did not ignore the importance of physiology, but he did not emphasize it to the satisfaction of the new academics. Instead Gray's primary interest in textbooks and in teaching was to prepare the student in plant identification and structural botany. He believed that when trained in structure the student would then have the basis for understanding plant functions and the relationships that existed between plants. The physiological and morphological material to which Bessey devoted two hundred pages — one-half of his book — Gray covered in twenty-eight pages. Bessey also placed much greater emphasis on studying botany in the laboratory using the compound microscope to investigate the minute internal structure of the plant. Gray, on the other hand, stressed more external structure which enabled the student using Gray's textbook to learn the parts of the plant necessary for its identification.[22]

Alphonso Wood and Almira Lincoln Phelps gave little attention to either structure or physiology, providing more of a simplified description of individual plants. Phelps and Wood were both secondary school teachers who had written popular books directed at beginning botany classes since the 1830s and 1840s. The wide range of intent and scientific knowledge contained in elementary textbooks revealed a major problem for American writers; that is, the beginning course for a student in botany could range from elementary school to college. Wood, and especially Phelps, lacked expertise when compared to Asa Gray, but the two popular writers had been very successful. Such writers increasingly became the special targets of academics. In his most recent text in 1879, for example, Wood teamed with J. Dorman Steele to write an elementary course of study, *Fourteen Weeks in Botany.* The reviewer in the *Botanical Gazette,* probably John M. Coulter, judged the book inaccurate and poorly conceived.[23]

Bessey also differed from his competitors by following the lead of Sachs in giving the nonflowering plants attention equal to that of the flowering types. Gray justified his own imbalance largely on pedagogical grounds. He believed that the vastness of the plant world would only confuse beginners if they tried to study it all. It was "best for the beginner, therefore, to treat higher orders of plants by themselves, without particular reference to the lower."[24] In his *First Lessons in Botany,* Gray included one chapter of twenty pages describing some of the families of the nonflowering plants, and in *Field, Forest, and Garden Botany* he devoted 450 pages to descriptions of the families of flowering plants but only a small fraction of that to the lower families. Four groups—the horsetails, ferns, club mosses, and selaginella—are the only nonflowering plants even mentioned. Aside from his belief that in textbooks and popular accounts of common plants the beginner could better understand and become interested in the higher forms, Gray's imbalanced treatment of the plant kingdom reflected his own interest in the flowering plants and in part his recognition that the "lower plants" were insufficiently understood.

By devoting greater attention to the "lower plants," Bessey demonstrated his interest in fungi, but more importantly, he recognized that German literature contained much more about these groups than many American botanists realized or were willing to admit. Bessey also believed that the "lower plants" would remain poorly understood as long as they were ignored in the classroom. The "lower plants" similarly held a key place in any hope of achieving a natural system of classification, he asserted. Most classification systems, Bessey explained, "kept alive the notion that so-called flowerless plants are quite different as to their reproductive organs from flowering ones," and the result has been, he continued, to create an "imaginary gulf between groups of plants which in nature are side by side." Improvements in classification such as going beyond external features and including the entire life of the plant, Bessey maintained, would remove much of the misunderstanding surrounding the nature of the lower groups.[25]

The reception of *Botany for High Schools and Colleges* was largely favorable and much of it enthusiastic. Bessey received at least twenty complimentary letters from botanists throughout the United States who for the most part praised the particular points that Bessey had hoped would make his book unique. Most writers stressed the " 'fundamental contradiction' between your [Bessey's] method of presenting botany to the learner and that of Gray."[26] J. C. Arthur, a recent graduate student of Bessey's, predicted that "your book cannot but exert a strong and salutary influence on the direction of botanical thought in this country." As a disciple of Bessey and of structural physiology, Arthur anticipated great changes in the study of American botany, and he saw the importance of the new textbook. "I look upon the advent of your book," he stated, "as the first rays of the dawn of a new era for American botany, when the name shall mean *botany* and not a superficial knowledge of the natural history and classification of flowering plants, and when to be an assiduous collector will not be sufficient to entitle one to the distinction of being a *botanist*."[27]

Other correspondents echoed Arthur's observation that the book was a landmark in American botany: "They [the two books in the Holt science series by zoologist A. S. Packard and Bessey] are by far the best yet produced in their respective subjects. . . . Professor Bessey was so much a stranger that I am surprised to see the masterly originality with which he handles his subject"; ". . . it is the Best Book of the Kind in the English language"; "I am delighted to see that much neglected part of the science of plants placed within reach of our students"; "[Your *Botany*] comes nearer the ideal textbook of botany for our American colleges than any other one yet offered"; and ". . . the book deserves all the good things that the reviewers have said about it, and it will, I am sure, take a leading position as 'the botanical text book' for advanced college classes on this side of the Atlantic."[28] Correspondents likewise congratulated Bessey for providing a boost to laboratory study with the textbook and for providing the American student with "the best results contained in the works of the great European masters."[29] These comments were accompanied with assurances that the textbook would be widely used in classrooms at such schools as Cornell University, Oberlin College, Alabama State A & M, and the state universities of Ohio, Iowa, Indiana, Michigan, and Maine.

The responses from Harvard were of particular significance to Bessey. Goodale had planned a physiological botany book for several years but had not yet finished it; he merely stated that he was "pleased" with Bessey's book.[30] Farlow was very flattering in his comments, and in regard to the parts on the "lower plants" he stated that it "seems to me that you have gone farther than any text-book in English and you have brought several points more nearly up to the times than is done in Sachs."[31]

Gray, probably not surprisingly for a competitor, was complimentary but generally had little to say about the new book. In a letter to Bessey, Gray acknowledged that the book was "very creditable," but he claimed that he had had "time only for a rapid glance at it."[32] Since Bessey's book represented a departure from the way Gray believed students must start the study of botany, it was doubtful that Gray approved of Bessey's overall approach.[33]

Gray was not alone in his apprehension that beginning students could not handle minute structure, function, and flowerless plants. The most frequent criticism of the textbook, even by botanists who otherwise liked it, was that the material was too difficult for high schools students and even for beginners in college. It was certainly not the general book Holt had wanted. Most persons stated that they planned to use the book in their advanced classes. Professor Eugene W. Hilgard of the University of California, for example, noted that the book was "valuable for the advanced student who wants to know more than names and morphology, yet I find few of this type student." He continued that he had "tried to keep classes interested in the details of vegetable anatomy and of microscopic life but found year after year dropping more to the view of Gray that it is first necessary to create proper interest of what the student can see."[34] Although Bessey claimed that he successfully used laboratory, microscopic, and nonflowering materials and methods with his own beginning classes, he found few other botanists at that time who believed such approaches possible or desirable.

Despite these major pedagogical differences, which were to remain well into the twentieth century, the formal reviews in the scientific journals likewise were favorable. Gray, who handled most of the botanical reviews for the *American Journal of Science and Arts,* took the time in the journal to comment on Bessey's book. Gray emphasized how overall the book testified to the progress taking place in American botany, and particularly he noted Bessey's contributions to the understanding of the "lower plants." He agreed that the book served to introduce the student to Sachs, and in regard to classification of the lower groups he considered Bessey's coverage an improvement. He tempered his praise, however, by noting that classification of the "lower plants" was the weakest part of Sachs. In conclusion, Gray judged the book a "commendable beginning."[35]

In the *American Naturalist,* which was beginning to challenge the *American Journal of Science* as the leading journal of natural history,

# A TIME OF PROMISE BUT UNCERTAINTY 31

A. S. Packard, Jr., one of the coeditors and author of the other book in the Holt science series, wrote the review himself rather than follow the more normal custom of turning it over to a "devotee of the flowering plants." Being a zoologist, Packard essentially used Bessey's preface to summarize the contents and uniqueness of the book. Agreeing that the coverage of the entire plant kingdom and the reliance on Sachs were the major contributions, Packard sided with Bessey's critics on the issue of the intended audience. Packard likewise believed that students should start with a familiar flowering plant and that this book was too difficult for most beginning students. He acknowledged, however, that the book was the best available for the teacher and for the college student who needed to have a general understanding of all plants and their physiology. "Without disparaging school books written by other botanists," Packard stated, "it seems to us that Professor Bessey's book is indispensable to the teacher of botany as it is or should be taught in these days in our leading colleges and universities."[36]

Bessey's textbook received its biggest boost when reviewed in the *Botanical Gazette*. The *Gazette* was a small, relatively new publication in its fifth year, edited by John M. Coulter who had recently moved from Hanover to Wabash College. Coulter wrote the review, giving it three of the twelve pages in the issue. Like Bessey, Coulter was devoted to bringing about changes in the study of botany; therefore, he was extremely receptive to Bessey's approach and asserted that Bessey had done justice to the "new botany." "You may be assured that I welcome this new departure as exactly the thing I want," he wrote.[37] Coulter stressed in his review that this textbook was not another Gray or Wood but "a new departure in American botanical text books." Tracing the past study of botany in the United States, Coulter noted that "to our country belong some of the very finest works on morphology and classification published and they rank as the very highest authorities, but physiology remains to be written." Under these circumstances Coulter acknowledged that little of the book could be original, but he did recognize the originality in the coverage of the "lower plants," particularly regarding classification. Coulter also was impressed with the suggestions Bessey included for laboratory work, stating that "this enables the student to go into the laboratory alone, or rather with the aid of the experience of Professor Bessey, one of the most successful of teachers, and perform satisfactorily all the elementary work in the histological structure of plants." In conclusion, Coulter commented that "we would most cordially commend the work to the use of all professors and students of botany as not only the *best* American book upon the subject, but the *only* one."[38]

The year before publication of his *Botany,* the State University of Iowa in recognizing his contributions to "science and to the state of Iowa" had awarded Bessey an honorary doctor of philosophy degree. Now the completion and warm reception of his textbook lifted Bessey from an ambitious, young teacher of local reputation to a botanist of national prominence.

At the same time he was writing his *Botany for High Schools and Colleges* in 1878–79, Bessey convinced Holt and Company to issue an American edition of an English textbook by William Ramsey McNab.[39] Bessey first had read the book in late 1878 and used many of the illustrations in his own book, and he judged that the McNab work was better in structural physiology than anything else available in the United States. The two volumes by McNab closely followed the general approach of Sachs, but Bessey believed both Sachs and McNab were too encyclopedic and technical for classroom purposes. In the future, Bessey intended to write a physiology textbook "on a very different plan," but for now time allowed only a rewriting of McNab.[40]

Bessey was most dissatisfied with McNab's presentation of reproduction and classification, and aside from the rewriting for the "wants of students," these sections were the only major revisions. The result was one volume that Bessey hoped would "direct attention to the study of plants as living things, rather than to their bare analysis and classification," a hope which persons close to Bessey were to hear stated repeatedly during his career. Although at first suggesting that the McNab book should appear in 1879 as a short text to precede his own *Botany,* Holt did not publish the revision until 1881. In comparison with his *Botany,* the revision of McNab garnered little notice in the journals or by Bessey's correspondents. During its first two years, the revision sold about half the number of copies of *Botany,* yet it still sold three times the fifty copies required for Bessey to receive his ten percent commission.[41]

Within four months after the publication of *Botany for High Schools and Colleges,* Holt and Company wanted a second edition. Although he asked many botanists around the country for suggestions and corrections, Bessey made only a few changes in either the second edition in 1881 or the third in 1883. He did update the citations and references to include recent literature in the fourth edition in 1885.[42]

His initial success with writing textbooks resulted in Bessey formulating very ambitious plans in the early 1880s for more books. He discussed with Holt and Company the need in America for another pocket manual that collectors could carry with them to provide descriptions of plant families and identification keys. A related need was a plant manual that covered more than the eastern United States and included more than flowering plants. Bessey indicated that these projects would have to wait, however, because he wanted to "push structure and physiological botany a good deal further in one or two more books first."[43]

As Bessey's ambitions outgrew his capacities, he likewise pushed aside the physiology texts. Both Bessey and Holt thought increasingly about a true beginning botany book. Obviously his *Botany for High Schools and Colleges* was not being used by teachers for beginning classes in secondary and common schools nor extensively in elementary college classes.[44] Difficulty of material was a problem, but the reluctance to use the book was more fundamental. Teachers were rejecting Bessey's approach of how to teach botany. Traditional teachers saw identifying flowering plants as a proper beginning because students would be more familiar with these plants. They also believed that physiology was inherently more difficult than taxonomy for the student and that beginners were not prepared to work on laboratory topics. In contrast, experience had convinced Bessey that using the laboratory, or learning by doing as he viewed it, was the best and easiest way to learn. In keeping with his view that botany should be the study of living plants and their forms and functions, Bessey maintained that study of the flowering plants was too difficult for the beginner because these plants were physiologically and structurally the most complex. Hence Bessey's idea of the "natural" way to study plants was the opposite of the traditional way; he wanted to start with the lower forms and progress to the most complex, the flowering plants.[45]

Criticism of Bessey's book reaffirmed the belief of Henry Holt that specialists in science made poor writers of elementary textbooks.[46] Again Bessey disagreed. As had Asa Gray, Bessey believed specialists should be the only writers of such texts. Textbooks then would be freer of the many errors Bessey believed filled most books and gave youngsters misconceptions about botany. Furthermore, specialists had the knowledge necessary to see the logical sequences of study and to accurately extract the fundamental points essential to building an understanding of science. The only weakness that Bessey saw in specialists was that too many did not consider carefully enough the "needs of the student" when preparing a course of study. Bessey won the argument with Holt and signed a contract in January 1884 to write another elementary botany textbook.[47]

In *The Essentials of Botany,* Bessey followed the general pattern of his earlier book, but he reduced the size and replaced most of the technical language with English terms. Yet it was not a simple duplication. Bessey eliminated altogether the topics that he judged were not basic principles or concepts, and he placed an even greater emphasis on laboratory work. This expansion of his views of how botany should be taught likewise pleased *Botanical Gazette* editor John M. Coulter. In his review, Coulter emphasized that while "essentially an abridgment," the changes were "so many and so important that it has the interest of a new work." Overall Coulter concluded that "no text book ever had a better promise of meeting a long felt want than this," and he agreed with the point Bessey had stressed that the book "will be welcomed wherever the aim is to learn from nature herself, and to make the book serve only as a *guide*."[48]

The *Essentials* was also favorably reviewed by George Goodale and by Joseph Rothrock, a pioneer in the use of experimental laboratories and new methods. Rothrock of the University of Pennsylvania agreed that an exclusively systematic approach limited the study of botany, and he noted how well this textbook fit the needs of the "new botany." He praised the style and method of presentation: "There is nothing superficial in it, nothing needless introduced, nothing essential left out." Particularly, he praised the guides for laboratory and microscopical work. Yet his only major criticism was that Bessey had gone too far by devoting only one chapter out of six to systematic botany and had ignored too much a discussion of how to analyze plants. "There is a danger," Rothrock warned, "that we may run from one extreme to another in our teaching, and in our new love for morphological botany, neglect systematic botany too much." He concluded that "a part never can be greater than the whole, and we regret the tendency, already too marked, of being content to study cells and cell growth and aggregation, without being able to name the plant on which the observations are made."[49] Interestingly, Bessey would later echo Rothrock's warning that botanists were neglecting plant identification, but now he was encouraging that neglect on the American side of the Atlantic.

In the late 1870s and early 1880s a variety of factors worked to keep Bessey's research and writings widely diffused. In addition to preparing textbooks he became editor of the botany section of the *American Naturalist* in 1880. Also, the frequently changing but always constant demands for agriculturally related botany kept Bessey moving from topic to topic. Bessey always complained of being too busy in a land-grant college setting to specialize to the degree desired, yet in the long run his knowledge of the basic literature, largely acquired at this time, and his being forced to keep abreast of the full range of botany proved advantageous. Within his chosen area of interest, he continued his simple observations and experiments in structural physiology, devoting particular attention to reproduction and leaf functions.[50]

Other interests that became evident at this time, again in conjunction with the writing of textbooks, were classification and evolution. Although Bessey never used the term *evolution* in any of his writings or addresses during the 1870s, he did discuss variation in plants and accepted the idea of natural selection. The few references he made to the topic at this time usually concerned Charles Darwin's *The Variation of Plants and Animals under Domestication*.[51] Bessey apparently was a confirmed Darwinian from the time of his undergraduate studies because he never mentioned any discussion, debate, doubt, or transformation having taken place in his thinking. His first mention of the subject was a matter-of-fact discussion that demonstrated his adherence to the idea of natural selection.[52]

By the 1880s Bessey frequently was commenting on the impact evolution must have on classification. He agreed with others on the subject that classification needed to "bring out the genetic relationship of the various groups which are considered to have descended from more primitive forms." The ultimate aim of the new stress on comparative plant structure was to disclose the relationships between different plant groups; therefore evolutionary theory demanded a "complete revision of classification." Appreciating the debt owed Darwin for popularizing this view, Bessey noted that "once granted the origin of species by natural selection and its full import understood and acknowledged, a mutual relation is seen to exist between one group and another, a relation which is much more than that of mere structural similarity." It was this perspective of plant development through modification that provided Bessey not only his conception of the plant kingdom but also his opinion regarding how one should teach botany. "Under the influence of the Darwinian method," Bessey observed, "the vegetable kingdom is assuming a shape in our classifications which shows a gradually increasing complexity, a gradual modification and differentiation as we pass from slime molds to flowering plants."[53]

What Bessey hoped to do in some of his own research was to trace the development of reproductive parts in "lower plants" and demonstrate their modification into true flowers in higher forms.[54] Although recognizing the difficulty of determining true genetic relationships between plant groups, he believed that comparative morphology provided the best chance for future success. Bessey also thought that the goal of natural classification could be achieved better than in the past by considering embryonic stages in addition to the customary reliance on structural similarity.[55]

In his textbook Bessey was able to place before the public his ideas of an improved classification system for algae and fungi, though this first venture received mixed and cautious responses in the reviews. In the

"Botanical Notes" of the *American Naturalist,* Bessey frequently commented on the general classification systems being developed by European systematists including George Bentham and Joseph Hooker, Julius von Sachs, Anton de Bary, and Adolf Engler, and on the classification of specific groups by Americans such as Gray, Sereno Watson, Edward Tuckerman, Lucien Underwood, Leo Lesquereux, and Thomas James.[56] More specifically, Bessey's own work on the Iowa flora allowed him an opportunity to organize a small segment of the plant world along improved systematic lines.

The state survey of Iowa plants still progressed slowly. Bessey published only one addition in the early 1880s, and this was a list of all the "lower plants" in the immediate vicinity of Ames with the rusts and smuts expanded to cover the entire state. This was merely a listing of species with the locations indicated for most.[57] J. C. Arthur, Bessey's former student, added several studies to the survey including his botany of Floyd County, Iowa, a *Catalogue of the Iowa Flora,* and a listing of the rusts of Iowa. With the intention of uniformity, Arthur followed the general classification in Bessey's textbook when possible, but in his list of rusts he stated that a "lack of knowledge of the relationships of the American species" forced him to resort to an alphabetical listing. In addition to their state survey of Iowa, Bessey and Arthur continued to collect and describe new species of fungi. Other botanists included these in monographs; particularly they contributed to the work of William Farlow at Harvard.[58]

By the early 1880s the accumulation of knowledge about American fungi was quickening as was Bessey's own expertise. Continuing to collect and identify new species in Iowa, he described in the *American Naturalist* some unusual and interesting forms of fungi and displayed a particular interest in the controversies over whether slime molds were plants or animals and over the nature of lichens. He sided with those botanists who placed slime molds on the animal side, and regarding lichens he seemed to favor the western naturalist George Englemann's concept of lichens as intermediates between algae and fungi. Bessey displayed his practicality when he stated that in both cases botanists were wasting much effort in argument when the distinctions were quite small and uncertain and the organisms existed regardless of how botanists chose to classify them. Although stressing to his readers that "lower plants" were still largely ignored, Bessey saw some improvement in the work of a few Americans such as Farlow, Arthur, Joseph T. Rothrock of Pennsylvania, Thomas J. Burrill of the University of Illinois, and another former student, Byron Halsted of Rutgers. Bessey offered additional praise for these few by stating that they were communicating their findings in "simple English."[59]

Closely related to his interests in fungi and practical botany, the nature of plant diseases became an area of increasing interest to Bessey. It was accepted by 1880 that fungi caused some diseases, or at least were associated with disease, but the nature and extent of this relationship was unclear. Bessey directed most of the time he spent on the subject revealing what was known about diseases, particularly that known in English and American literature. Mostly he related what was not known. In 1882 he emphasized that there was still no written work on the subject in the English language other than a few isolated, specific studies. "It does not speak well for the kind of botany we have been teaching in this country for the last fifty years," he noted, "that no attempt has been made in all this time to give us a treatise on the diseases to which plants are subject." The few general articles available were in German and French, and little of this information had been absorbed by Americans. In particular, he pointed out, as with most areas of science there was little known about the plant diseases of the American Midwest and West.[60]

In an attempt to fill this void, Bessey continued compiling lists and descriptions of the various harmful fungi of Iowa. Again Bessey combined the scant but growing body of scientific knowledge with his specific investigations in Iowa. Bessey wrote numerous popular articles in which he described in depth troublesome fungi for Iowa farmers. In these writings he provided detailed descriptions of the harmful plants including illustrations, where they grew, the type of disease or injury inflicted, descriptions of the diseases, and suggestions of how to rid the farm of the troublemakers.[61] In two of these articles for the Iowa State Horticultural Society he abandoned his normal listing of species and discussed in plain terms plant diseases and harmful and nonharmful fungi. The first article explained how fungi could cause injury and disease, and the second discussed plant diseases in general. Both were good introductions to the subject of pathology, but again Bessey had to admit to his agricultural audience that this was as far as science had progressed.[62]

In addition Bessey tried to explain to botanists and agriculturists what scientists needed to learn from their investigations before botany could be of much benefit to farming. Basically botanists needed to expand their general understanding of the nature of parasitic fungi, particularly the complicated life cycles involving secondary hosts, which were little understood at that time.[63] Hope for improvement received a tremendous boost in the early 1880s when Thomas Burrill and J. C. Arthur opened a new, virtually unknown area of research by attributing a plant disease to bacteria.[64] Bessey corresponded directly with Burrill to learn as much as he could about the research. While remaining cautious and apprehensive about the findings at first, by 1885 — and probably sooner — Bessey had accepted the germ theory of disease in regard to plants and exhorted botanists to turn their attention to the subject.[65]

Bessey's activities in practical science continued to revolve primarily around the Iowa State Horticultural Society. In 1880 the horticulturists rewarded him for his contributions by naming him, along with President Welch, one of eleven honorary life members. He presented a paper at each annual meeting from 1878 through 1884 and was an active participant in discussions. Desiring to introduce new plants into the state, the society was particularly attracted to Bessey's work on the climatic adaptation of plants. This differed somewhat from his earlier work concerning climatic effects on the physiological functioning of the plant and was more a general consideration of plant geography.[66] While at Harvard in 1875 he had obtained evergreen seeds from the Rocky Mountains. When proposing that these seeds could be grown successfully in Iowa, his audience in the Eastern Iowa Horticultural Society objected, stating that his ideas did not match with their experiences in tree and fruit growing in Iowa.[67]

Bessey continued to investigate the subject and presented his general ideas on climatic adaptation to the state society in 1878.[68] Explaining how plants came to occupy definite regions and how American flora were affected by the still-disputed glacial epoch of their geological history, he concentrated on the factors of temperature and moisture. When discussing how plants adapt to new environmental conditions and localities, he relied heavily on the work of British naturalist Alfred Russel Wallace and also quoted from Herbert Spencer's *Principles of Biology* and Charles Darwin's *The Variation of Plants and Animals under Domestication*. Bessey concluded that "all or nearly all species vary" and that the variations "induce constitutional differences which render some of the varieties capable of withstanding greater amounts of heat or cold, or more or less of moisture." As horticulturists, Bessey observed, the society members were all familiar with the adaptation of cultivated plants, but "hitherto in too many cases the improvement has been more or less accidental. In the future we shall look more and more for intelligent effort directed to a certain purpose."[69]

Although the society directed most of its efforts to importing plants from northern Europe, Bessey and society member C. C. Parry, amid continued objections, maintained that American plants—particularly conifers from the Rocky Mountains—could survive the dry, cold winters of Iowa. The need for trees and the relatively cheaper acquisition from the Rockies helped Bessey convince the society to ask the state legislature to finance acquiring and testing the hardiness of possible plant introductions. Encouraged by a small appropriation to purchase the Rocky Mountain conifers, the society next lobbied the state to provide land on the college farm for an experiment station where systematic testing of new plants would replace the sporadic efforts of individual horticulturists. The state approved the plan in January 1878. By 1880 the horticultural station, under the direction of J. L. Budd, had plants from as far away as Russia, northern China, and Sweden growing on campus. The one hundred or so dollars that the Iowa State Horticultural Society provided each year for the station, however, was an insufficient investment to carry on the variety of tests desired by the state society, and again its members turned to the state legislature for additional money. This time they were unsuccessful. The college leaders likewise claimed that they were so short of funding that they were unable to provide more than the land and the time of several of their professors for the station. It was such experiences as these that convinced groups like the State Horticultural Society to turn to the national government for more substantial and permanent funding of agricultural experimentation.[70]

Although not directly involved with the horticultural station, Bessey did offer the services of his laboratory and invited members of the horticultural society to submit questions and problems to him.[71] Like Bessey, President Welch continued his active role in the State Horticultural Society. In addition to writing papers and doing committee work, Welch occasionally addressed the group on the nature of scientific knowledge and on the relationship of science to practical education.[72]

In these ways the faculty continued their ties with the State Horticultural Society, and for their part, members of the society continued to side with the Welch group. They supported scientific agriculture, even though not all the members agreed on what that meant, and backed efforts of the college to obtain more funding from the legislature. Criticism of the Welch administration had never ceased, and the issue of the intention of practical education and the nature of the land-grant college was certainly not solved or a dead issue in Iowa by the 1880s. Yet in the horticultural society the Welch group had nurtured an important ally.

The Iowa State Agricultural Society, however, ceased to attract Bessey's direct involvement during the late 1870s and early 1880s. He attended only one meeting of the society during this period, but it provided some interesting discussion. Seaman A. Knapp, the new professor of agriculture at Ames, addressed the society on the proper type of training at the land-grant colleges. Knapp advocated a practical education with enough "liberal instruction as would preclude narrowness." Elaborating upon what he meant, Knapp said he favored teaching science but not the same type as would be taught in the traditional liberal arts. Likewise he asserted that the Morrill Act intended experimentation to be a "prominent feature" of agricultural colleges. But he did not see the laboratory as a replacement for the model farm, and he wanted practical farm work taught equally with practical science. He also advocated requiring manual labor of all students as part of their training.[73]

In the discussion following the address, there was much concern among the members as to whether practical instructors at the college such as Knapp and J. L. Budd were isolated among the other faculty. Knapp did not believe so, stating that the method of teaching was really what made science practical or not, and he used botany for his example: "It is called science to teach them [students] the names of the plants, so that if you should hold a plant up, they could tell you the name and order of it and so on. Now, so far as the farmer is concerned, it don't amount to a snap of the finger." Knapp continued that what was important was to know how plants grow and the usefulness of the plant. "To study such questions is another kind of botany," he noted, and concluded that "if our professor of botany endorses that theory, he is just as good an agriculturist as I am. And he does endorse it."[74]

And indeed Bessey did agree with Knapp's analogy. When finished with his remarks, Knapp called upon Bessey for further comments. Bessey was pleased that he and Knapp so thoroughly agreed upon agricultural education. So different was the old science from the new, he added, that "I do not like that old word botany." He then reminded his audience that although younger than most present he probably had more experience with agricultural education, having gone from a farm to the "mother of agricultural colleges" and then to Ames. Particularly, he noted, he had a long association with manual labor education. Again he believed he and Knapp completely agreed on the need for an "out of doors system." Here Bessey seemed to ignore the differences that existed between them. At this time Knapp apparently favored manual labor for the traditional reasons that it encouraged hard work, was healthy, provided practical training, and separated the working-class student from the effete upper class. Both recognized that manual labor had been modified by the land-grant colleges during the past decade and that it did not have to be the "how to plow or milk a cow" type of training. But Bessey believed that field trips and laboratory studies could replace farm work entirely and still be an "out of doors system." It was not clear that Knapp would go this far. While convinced that the way he taught botany was practical, Bessey also believed it was the best way to teach all students, not just agricultural students. Because students developed more understanding of the "plant as a living thing," he argued, their knowledge was more useful regardless of whether they intended to apply it to agriculture. Despite the similarity of their views on education, Bessey and Knapp had more differences than they perceived or were willing to admit to their agricultural friends.[75]

In addition to the state institutions, the national scene increasingly drew Bessey's agricultural interests—particularly the U.S. Department of Agriculture (USDA), where important developments were underway. Bessey, who liked to write to government officials, emphasized to the president, members of congress, and officials of the USDA that the history of the department was often "far from creditable," yet potentially it was the pillar upon which agricultural science could build in the United States.[76]

The first twenty years of the USDA had not been glorious. Scientists had become disillusioned with the department and the department with the scientists. Farmers and livestock producers had little reason to look to either the scientists or the department for any substantial help. Yet by the early 1880s efforts seemed underway to expand the practical scope of science and education and potentially to change the nature of agriculture. Commissioner of Agriculture George B. Loring initiated a series of national agricultural conventions in the early 1880s that brought small groups of professors and experimenters to the nation's capital to discuss problems of agriculture. The intent of these conventions was to start the small USDA on its way to becoming a useful servant to American farmers, gardeners, and ranchers. Amid the numerous meetings and discussions, the conventioneers agreed that many agricultural principles were not formulated or settled, that agriculture needed unified long-range research conducted throughout the country, and that scientific research was as much the duty of agricultural colleges as was teaching.[77]

A prominent theme of the conventions was that agriculture needed the aid of experimental science. This theme was not new among American agriculturists, but the decade of the eighties marked the beginnings of concerted group efforts to promote agricultural science and to infuse it into the governments and colleges of the nation. There was no clearer expression of this effort than the forming of the Society for the Promotion of Agricultural Science (SPAS). In 1879 in an editorial, "A Plea for Agricultural Science," E. Lewis Sturtevant of the New York Experiment Station and a leading promoter of the organization had stressed that scientific farming in America hardly went beyond soil analysis and that what was needed was a society composed only of scientists who would publish papers that were truly contributions to knowledge. Restricting the membership to experimenters and limiting the number so as to make it a select group, twenty-one members of the society in 1880 dedicated themselves to applying more science to agriculture and to making the study of agriculture respectable among scientists.[78]

Sturtevant, William J. Beal of Michigan Agricultural College, and George C. Caldwell of Cornell University had initiated the actual organization of the SPAS with a circular letter to fifteen scientists who had demonstrated ability and interest in agriculture. They invited the group to gather at the 1879 meeting of the American Association for the Advancement of Science (AAAS) to form what they hoped would "come to be regarded as the authority in the field that it occupies as does the National Academy of Sciences in its special domain." They also hoped this organization would supplement, in the scientific realm, the Association of American Agricultural Colleges.[79] At the first formal meeting in 1880, twelve of the original twenty-one members attended. Beal became the first president, followed the next year by W. H. Brewer of Yale University. Early meetings typically had ten to twelve members present who read about fifteen papers ranging from seed and milk testing to agricultural education. Membership remained small due to the select nature of the group, some disinterest, and a rule that dropped members who failed to attend and read a paper within a three-year period.[80]

Bessey had been active enough in agriculture to be among the original fifteen scientists contacted. He became a founding member of the SPAS although he was certainly not one of the driving forces in the beginning. He did not attend his first meeting until 1883 when its proximity in Minneapolis and a reminder by Beal of the three-year rule prompted his appearance. However, relative isolation and lack of travel money explain his inactivity more than lack of interest. In its early years, Bessey's biggest boost to the SPAS and its desire for scientific status came in the form of an increased amount of attention to its work in the botany section of the *American Naturalist*.[81]

A final area of agricultural activity by Bessey at this time was his continued commitment to popular writing and activities. The college leadership had attempted to establish and maintain contact with the public by encouraging the faculty to participate in agricultural fairs, to write articles, and to answer questions from correspondents in the agricultural press. Starting in the mid–1870s, the Iowa college entered into cooperation with and occasionally operated the agricultural journals *Progressive Farmer, The Producer, Western Stock Journal and Farmer,* and the *College Quarterly.* By the early 1880s Bessey also began writing for the *American Agriculturist,* the outstanding agricultural journal of the Midwest, and contributed a series of articles to the *New York Tribune* and the *Chicago Herald.*[82]

Bessey's stay at Iowa Agricultural College ended in 1884. He confided that he moved to accept a better offer and opportunity. A number of small irritants and displeasures never were resolved, and the opponents of the Welch administration remained active. Among the irritants, low pay was a constant. Applicants for jobs at Ames made Bessey aware that even in a day of low salaries for professors, Iowa Agricultural College did not pay well. To compound the problem, the board lowered salaries in 1878, forcing Bessey to give up his house in Ames and move onto the campus. Bessey felt that the job itself was becoming increasingly diffuse and burdensome. He had tried constantly to avoid teaching nonbotanical courses, but he was not free of the last of these, zoology, until 1880. He wittily referred to this situation as "occupying an elongated 'settee' instead of a chair" in botany. Time spent on correspondence increased significantly during the early 1880s as more people looked to the agricultural college for guidance as Bessey had encouraged them to do. He also increased his popular writings in the agricultural press. Educational groups more frequently asked Bessey to speak, and in 1881 he was appointed to the advisory committee of the State Superintendent of Public Instruction for Iowa as representative of industrial training. Within the college Bessey received a heavier administrative load. In his early years at the college, Bessey had served as secretary of the faculty, an undesirable position apparently given to newcomers, but the board elected him acting president in 1882 and vice president "without extra pay" in November 1883, a position Bessey filled for one year.[83]

This situation was occurring at the same time Bessey was gaining greater national recognition. He found the many—and seemingly increasing—duties that his job entailed restricting his opportunities as a botanist. Bessey kept his complaints low-key, but ever since his failure to secure the position at the University of California in 1875–76 he stated that the actions of the Iowa board discouraged him and indirectly indicated a desire to leave.[84]

Following the legislative investigation in 1874 Bessey and other members of the Welch group had instituted curricular reforms in which they pushed the teaching of the general sciences ahead of practical or vocational teaching. This was in line with their views concerning the need for scientific agriculture. Opposition continued, however, and in 1880 President Welch chose Professor Seaman A. Knapp to take charge of the agricultural teaching in an effort to satisfy those who still accused the college of not being practical enough. Knapp, a former livestock producer and editor of an agricultural journal, was a proponent of scientific agriculture but only a recent convert. As indicated in his discussion on education with Bessey before the State Agricultural Society, however, there were important differences between them. Knapp increased the agricultural offerings at Ames and emphasized more conventional, practical training. He divided agricultural students and science students by creating a separate degree, the bachelor of scientific agriculture, a development Bessey opposed.[85]

Nevertheless the president and members of the faculty were still forced to explain what the college was attempting to do and why. Their speeches demonstrated the difficulty of defining the exact issues and disagreements that existed on the question of agricultural education. Occasional articles and letters in the local newspapers likewise revealed the fact that supporters on both sides, and many of those on the same side, had difficulty specifying their differences.

In 1881, for example, speaking before the friendly State Horticultural Society, President Welch made another attempt to clarify his views about industrial education and how Iowa Agricultural College was adhering to these views.[86] Still defending the central position of the basic sciences, he emphasized that such training was practical because it had changed the nature of traditional science education. "In fact," he stated, "while the progress that is not based in a measure on science is without profit, the science which does not embody itself in useful things is an empty show. In industrial education it is our policy, therefore, to close up everywhere the gap between science and practice." He then concluded that "we have carpenters, blacksmiths, ordinary farmers, and mechanics enough. Nineteen in every twenty of the whole population are given to industrial life. The great crying want is for master workmen. That science alone can furnish the means for the training proposed it is needless to say."[87]

Despite such attempts at persuasion, the smoldering dispute over the nature of the college and the leadership of Welch broke into open conflagration again in the fall of 1882. An opportunity for the opposition to make another attack occurred when the USDA appointed President Welch to represent it on a tour of European agricultural schools. The Iowa board of trustees first denied Welch a leave of absence but finally consented, leaving Vice President General James L. Geddes in charge of the college. In July 1882 the board met and decided it was best not to have Geddes, who also was the college treasurer, keep both positions. The trustees removed him as vice president and replaced him with Bessey. At their November meeting the anti-Welch element of the board was able to obtain a reduction in pay for Welch and for his wife, who also was a member of the faculty. In addition they removed Geddes from all his teaching and administrative positions and changed the teaching assignment of another faculty member of the Welch group. Although there is no record of his motive, Bessey resigned as acting president at this same meeting. Knapp, who was more acceptable to the agriculturists, was elected to replace him. During the next year while Welch remained in Europe the opposition, working through the Grange with its press backing and through certain unidentified members of the faculty, pressured the board for more changes. In November 1883 the trustees—by the vote of three to two—fired Welch as president and promoted Knapp to the position permanently.[88]

An immediate uproar ensued in which the board members on each side defended their actions, charges and countercharges blasted from both sides, and repercussions of the action drew comments from the press throughout the state. The stated motives behind the dismissal were confusing and conflicting, but in general the opposition charged that Welch had not provided the active leadership necessary for the advancement of the college since a breakdown in his health in 1877, that he was a poor teacher and administrator, that religiously he was a bad example for students, and—most importantly—that he "was not a true champion of industrial education."[89]

The defenders of Welch and his approach to practical education came immediately to his aid. Supporters portrayed the three trustees who voted against Welch in a variety of uncomplimentary ways, including "revolutionary," displaying "a high-handed exhibition of personal spite," "blind" in their power, "unmitigated cranks, utterly unfit for the duties laid upon them," "ignorant and incompetent," "manipulated," "men of little learning and little reputation," and "weak minded politicians."[90]

Bessey agreed with these latter opinions, stating that the actions reflected "jealousy, rivalry, and religious fanaticism" by supporters of "educational heresies." In contrast to Welch, a "scholar who could see far into the future, who could see that here must be built a real college on a foundation as broad as its charter," Bessey continued, the antagonists "were those who clamored for a cheap 'quick meal' type of school, which might appeal to the ignorant and the uninformed, and for whose support class prejudice might be arrayed."[91] Although one of Bessey's strengths in agricultural education was his ability to mingle easily with practitioners, he was most at ease with agricultural leaders, and he certainly had a vein of intellectual elitism that saw a part of education as developing middle class "gentlemen" apart from the common populace.

The uproar concerning the condition of the agricultural college carried over into the next session of the state legislature in March 1884. Welch had strong support. The entire faculty, including Knapp, formally requested Welch's retention, and a petition signed by one hundred fifty residents of Ames supported him. Some of the old, long-time sponsors of the college and members of the state societies also supported Welch. To many observers of the controversy surrounding Welch, one problem with the operation of the college was that the small size of the board of trustees — five members — enabled a small faction to gain control and act against the wishes of most persons involved with the college. The legislature responded by enlarging the board and by formulating a new guideline for the agricultural college that consisted of a broader, more liberal interpretation of the intent of the Morrill Act, both moves aimed against the anti-Welch faction.[92]

The clamor surrounding the college continued through 1884. Although Knapp seemed uninvolved in the removal of Welch, the anti-Welch group looked to Knapp as their champion and the Welch group refused to accept him, thus leaving the new president caught in the middle. In the midst of an enrollment decline during the year, concerted efforts continued among most of the faculty to have Welch restored. Finally in December 1884 the new, enlarged board moved to rectify the actions of the previous year. They removed Knapp from the presidency but retained him as professor of agriculture, reappointed Welch to the faculty, and tried to readjust some of the faculty and administrative turnovers, as with Geddes. Although these efforts were an attempt to find a middle ground, the new board was unable to settle the college's problems of administrative leadership during the remainder of the 1880s.[93]

In June 1884 in the midst of this furor, the University of Nebraska, a neighboring land-grant school, offered Bessey the chair of botany. After a visit to the campus where he found the botany department virtually nonexistent, Bessey rejected the offer. In August, however, he accepted when another offer came from Nebraska that included a deanship and "a very considerable increase in salary and the promise of appropriations for building up the department." His actual move did not occur until November 1884.[94]

In 1890 Bessey returned to Ames to help celebrate the quarter-centennial of Iowa Agricultural College. In his address, aptly titled, "Laying the Foundations," he reminisced about his experiences and memories of those early years of the college that he had witnessed from 1870 to 1884, a time that indeed was basic to the land-grant movement, to agricultural science, and to Charles E. Bessey.

# 3

# New Foundations: The Move to Nebraska

IN LINCOLN, NEBRASKA, Charles Bessey found a university in much the same condition as that of Iowa Agricultural College a decade earlier. The University of Nebraska had operated since 1871 but from Bessey's viewpoint it had made little headway in the areas of pure or applied science. Bessey officially did not assume his duties as professor of botany and horticulture and dean of the Industrial College until January 1885, but his involvement in the university began the previous fall.

Chancellor Irving J. Manatt recommended to the state superintendent of public instruction that Bessey represent the university and arrange an exhibit of the flora and fauna of Nebraska for the World's Industrial and Cotton Centennial Exposition. Bessey received a ten-day trip to New Orleans for his efforts, and in organizing his part of the exhibit he became acquainted with a number of educators and agricultural leaders from his newly adopted state. In particular he met and worked with Robert W. Furnas, a politician and longtime leader of the State Board of Agriculture, who was to become closely associated with Bessey in Nebraska.[1]

A more important introduction to Nebraskans and the university was Bessey's inaugural address in September 1884 as the new dean. In the address, aptly entitled "Science and Practice," Bessey outlined the ideas that had become so familiar to Iowans over the past decade regarding the "new botany," the purpose of the Morrill Act, and the role of practical science in a land-grant school.

Using horticulture as his example, Bessey discussed a variety of ways that the basic sciences of botany, chemistry, physics, geology, meteorology, and entomology contributed to improved gardening. Able only to outline the ideas that the Welch group had been striving to achieve in Ames, Bessey then explained what he believed should be the duties of the

Industrial College he was to head. The college, he indicated, should do what it can do best and what is not being done elsewhere—provide instruction in the basic sciences and their applications. "Can we safely offer more?" he queried. "An experience of many years leads me to the conclusion that we cannot." Bessey definitely wanted to steer the college clear of the "how to plow" type of vocational education. He stated, "The college can tell how plants grow, but it is not within its province to tell what plants to grow. The former is general, and true for all places and under all conditions, the latter is local and is modified by a thousand contingencies." He continued, explaining that the "college with its one, two or three men cannot hope to compete in the discovery of new methods of practice with the thousands of quick-witted gardeners whose livelihood and financial prosperity depend upon the discovery of better methods." Bessey concluded by emphasizing that much of the "odium" directed at land-grant colleges for their "loose instruction" could be avoided if the colleges kept to what they were most capable of doing, which was teaching general scientific information.[2]

Regarding the new botany department he intended to establish, Bessey informed his audience that the "day long since went by when the scientific botanist was one who collected, dried and pressed into dead flatness the plants of his neighborhood, only to attach to them afterward certain Latin names, by which they were thenceforth to be designated." For the first of many times to come, the students and faculty at Lincoln were told that "a plant is a living thing." The University of Nebraska in establishing a department of botany for the first time was not, Bessey charged, to be among those colleges that "still regard botany as a pleasant pastime consisting mainly of flower hunting, a fit thing for sentimental girls and effeminate young men, but unworthy the time and attention of strong minds." Bessey then explained the "new botany" in sufficient detail to demonstrate to his audience his intense devotion to the science.

Botany had been taught at Lincoln as part of a course in natural history, a common practice but one that was giving way to specialization. In his new setting there were no microscopes or any other laboratory equipment, but part of the agreement that lured Bessey to Nebraska was the promise of money for these purposes. In December 1884 the new dean gave his first lesson to the regents regarding the need for laboratory as well as field work in science. The need for a "personal acquaintance with nature" required not only extensive work in the laboratory, but also the use of such aids as a botanical garden, an herbarium, forage and grain gardens, and a museum with display items such as a woods-of-the-world collection, dried crop plants, and harmful fungi. There were none of these in Lincoln.[3]

True to their word the regents provided $5,000, and during his first year Bessey ordered his first six microscopes and other necessary equipment and offered laboratory classes in anatomy and physiology. More microscopes and accessories soon followed. The summer of 1885 saw the botanical garden started. Regarding the herbarium, Bessey had his personal collection, but one of his regrets in moving from Iowa was having to leave an herbarium that he had worked so hard to build. In starting over he purchased specimens from professional collectors like C. C. Parry, J. G. Lemmon, and A. H. Curtiss. Yet with little to spend on plant collections he also had to find ways to obtain free plants. Bessey immediately wrote U.S. Commissioner of Agriculture George Loring and asked him to contribute duplicate plants from the Division of Botany. These donations became a major part of the fledgling Nebraska herbarium. Also, as he had done in Iowa, Bessey quickly acquired correspondents around the state who collected and sent specimens to the university.[4]

Within a few years Bessey had the basic equipment for the type of botanical instruction he believed the "new botany" demanded, and in 1888 his efforts were further rewarded when a third building was added to the Lincoln campus. For use by the Industrial College, the new building provided additional space necessary for the expanding botany department, including a special laboratory for plant physiology. Further evidence of Bessey's presence was the cornerstone of the new building with its inscription, "Science with Practice."[5] During the establishment of his new department Bessey kept his colleagues around the country informed about the methods and equipment he found valuable in his new laboratory setting. His laboratory certainly proved to be one of the best models for the increasing number of such facilities that were developing in the United States.[6]

In his larger task of heading the Industrial College, Bessey was not starting from the beginning as he was with the botany department, but nevertheless he faced the momentous task of molding the college to his ideas of practical science. The University of Nebraska had had an Agricultural College since 1872 and an engineering course since the creation of the Industrial College in 1877, but the program had not been successful in gaining students or public support. As Bessey had found in Iowa, there had been constant disagreement over the intent of the Morrill Act between the vocationalists or systematists and those who wanted something in the way of scientific training. A situation not faced in Iowa was the attachment of the Agricultural College to the state university.[7]

To complicate the situation, Bessey came to Nebraska in the midst of an administrative and faculty shake-up. In 1882 the regents had removed the chancellor and three faculty members, and the next two years saw three more faculty resign, including Bessey's predecessor S. R. Thompson, professor of agriculture and dean of the Industrial College. The regents had charged Thompson with incompetency. In January 1884 a committee of the State Board of Agriculture had investigated the college farm, found "more theory than practice" with the farm in disrepair, and in general judged the college incapable of providing an agricultural education. The committee charged that the college and farm could never be adequate as long as they remained attached to the state university headed by regents who had no interest in agriculture. The committee's recommendation to create a separate college, as in Iowa and some other states, in which the university community would not downgrade practical education or view it disparagingly resulted in the introduction of a bill into the Nebraska legislature in 1885 to achieve the separation. In part the move was an attempt to increase the position and appropriations of the State Board of Agriculture, and in part the separation movement gained momentum through the efforts of some of the recently dismissed faculty to gain revenge against the regents. But the major concern of the separatists was that the university was using money intended for agriculture to support work in traditional arts and sciences.[8]

Some of Bessey's ideas about scientific agriculture thus were bound to meet resistance in Nebraska. In his first report to the regents, Bessey repeated the proposals in his inaugural address about practical education and likewise discussed them with the State Board of Agriculture. Bessey wanted to train scientists; therefore, he wanted the industrial students in regular science courses, not in some simplified courses geared to vocational training. He wanted the industrial students to be a part of the regular university, meeting the same standards, taking the same beginning courses, and specializing in practical scientific applications only at the end of their college training. In opposing the separation of the Industrial College, he feared the practitioners would gain control of the curriculum and any hope of academic respectability would be lost.[9]

With the vocationalists' power largely outside the university, Dean Bessey had little trouble convincing the chancellor and regents to implement immediately many of his desired changes in the Industrial College. Neither of the courses of study in his college—the agricultural nor the engineering—currently led to a bachelor of science degree, which Bessey wanted for industrial students. Unable to obtain the B.S. degree for his college until 1889, Bessey first concentrated on making the current four-year degree programs more general and more scientific. To train practical scientists, Bessey put the industrial students in the same basic science courses as the arts and sciences students, required extensive language training necessary to keep abreast of foreign research, included cultural courses such as history and literature which any "true university experience" needed, stressed extensive laboratory and field work instead of manual labor and lecture courses, provided applied scientific courses in the junior and senior years, and introduced, at least in the agricultural course, a senior project of original research.[10]

The scientific course of study brought academic respectability to the Industrial College but it did not draw students or acceptance by most agriculturists. The immediate result was to drive away the ten to twelve agricultural students currently connected with the college farm. As Bessey later remembered, "Not one was able to pass the somewhat stiffened requirements that were soon inaugurated and every one of them disappeared. They were the fellows who could not get along in any other course than the agricultural and horticultural, and even in this work they failed when the work was made a little harder." But Bessey never doubted the wisdom of his program: "As a result the University for a time had not a student in agriculture and horticulture, but when they began to appear again they were boys of different stuff."[11]

Despite Bessey's claim that the new students were better, they were no more numerous than before. Until 1889 the college never had more than fifteen students and most of them were in engineering. The chancellor, regents, and other supporters of the changes to practical science could only boast that industrial students were now academically equal to other students in the university and that the standards of the Industrial College were the highest in the country, but they could not silence the constant critics who still believed the university was not conforming to the requirements of the Morrill Act.[12]

The new dean worked hard to convince his critics that what the Industrial College was attempting was correct. As demonstrated in Iowa, Bessey was not opposed to practical vocational training, he just did not believe it belonged within the university, and there was a two-year course in agriculture aimed at improving the practitioner. The personnel of the university, particularly those connected with the Industrial College, he believed, had an obligation to take the findings of science to the agriculturists and in so doing demonstrate the value of science.

The university had sponsored Farmers' Institutes in Nebraska as early as the winter of 1873–74, but apparently they were not regularly maintained or vigorously supported until some of the county farmers' organizations attempted to revive them in the early 1880s. Bessey began participation in these meetings as early as December 1884 when he encouraged the agricultural society of Lancaster County to sponsor a Farmers' Institute with his help and that of his faculty. Enough institutes were held during the next few years that Bessey finally asked the regents to devise a more systematic method of involving more faculty so that he and H. H. Wing, professor of agriculture, would not have to attend them all. In 1888 in cooperation with Former Governor Robert W. Furnas, Bessey formalized a speakers list for the institutes and obtained funding, moves that greatly stimulated and increased the number of such meetings.[13]

Bessey also tried to popularize the Industrial College by placing advertisements in the farm journals, but still the students did not come. Further to create a receptive audience for his ideas, Bessey contacted numerous organizations during his early years in Nebraska. He became particularly active in the State Board of Agriculture. The board had long displayed an interest in science, and a part of each annual report contained essays on geology, entomology, zoology, meteorology, and botany. In 1885 Bessey became the state botanist for the board and in that year started a series of articles on grasses and forage plants. He also took an active role in preparing exhibits for the state fair, which was the major function of the state board.[14]

As in Iowa, Bessey became a regular participant in the State Horticultural Society.[15] In addition, during his first years in Nebraska he delivered addresses and papers to a variety of other organizations such as the Nebraska Dairyman's Association, the Stock-Breeders and Short-Horn Breeders Association, the Nebraska Improved Stock Breeders Association, and the Nebraska Convention of Stockmen.[16]

Closely related to these activities, Bessey wrote several series of articles and notes for two of the leading agricultural publications, the *American Agriculturist* and the *Breeder's Gazette,* and eventually became part of the regular staff of contributors of both.[17] Through these efforts Bessey's name and the work of the Industrial College became widely known throughout the agricultural community of Nebraska. Still, although his popularity and reputation increased and won support for the university, Bessey's accomplishments never completely silenced the critics of his reform of the Industrial College.

Bessey believed another way to convert critics to his concept of scientific education was to demonstrate the value of agricultural research. To do this, in light of the admitted meager results of agricultural science thus far, the Industrial College needed to become research oriented. The Agricultural and Industrial Colleges at the University of

Nebraska had engaged in experimentation since the early 1870s.[18] Most of it, however, had been simple observations similar to that done by Bessey, Budd, and McAfee in Iowa and similar to that done by most agriculturists and horticulturists around the country. The findings were few, the work scattered, and the results very limited in application.[19]

In Bessey's opinion agriculture would improve only when the applicable sciences were advanced and when new knowledge was unearthed and applied. Under current conditions applied science could never hope to attract the type of scientists needed until it adopted and attained scientific respectability. To obtain or develop competent applied scientists, his Industrial College needed the respectability necessary to attract the best students coming to the university, and applied studies had to offer an academic challenge and the promise of professional rewards for both the faculty and students.[20]

During this period in the United States the "cult of science" was continuing to catch hold of the public, and its promised rewards and results were appealing. Most scientists, however, desired to create an elite group of researchers, and while the improvement of practitioners was part of the promised rewards of the new science, these rewards were more indirect and in the future. The persons who wanted to separate the Industrial College from the university could easily, and correctly, regard the move by Bessey to more basic science and an upgraded curriculum as a further guise to promote an elite education and to change the interests and future of farmers' children into something other than merely to be better farmers.

In his initial report to the university regents in December 1884 Bessey discussed the research program that the Industrial College should pursue. He explained that there were two distinct "classes of experiments and observations": (1) the popular class "which aim to reach immediate results," and (2) the scientific "in which the aim is to discover some profound principle, or establish beyond dispute some fact in nature." "It would be well," he continued, "to draw a sharp line of demarcation between these two classes of experiments and then to adopt a fixed policy with respect to each." The popular experiments were important and the college should pursue some of them, Bessey acknowledged, but he emphasized that the results of such work were generally more immediate in time and confined to a small geographic area. Also, popular experiments constantly were being conducted outside the university setting. Scientific experiments, on the other hand, were more important because of their general nature and hence their more universal usefulness. In comparing the relative importance of the two areas of research, Bessey quoted from an editor of *Science* that the "great need of agriculture to-day is not new varieties of plants or improved breeds of animals, new methods of cultivating the soil or improved systems of farming; all these, and many other like things, are good; but the two great wants are a better knowledge of principles and greater intelligence to apply them."[21] Scientific work, Bessey continued, required special training and equipment that individuals and private businesses usually lacked. Therefore, he concluded, it is "especially important that experiment-stations and colleges which have the facilities for such experiments should be encouraged and supported in undertaking them to as great an extent as may appear practicable in each particular case."[22] Although some of the land-grant colleges had persons trained for such work, few had the equipment and time necessary to conduct research successfully.

Another serious limitation to agricultural research in the 1880s, as Bessey pointed out to the regents and to his fellow botanists, was that most scientists were not doing the type of research—that is, physiological and pathological botany—that would produce principles upon which agriculture could draw for popular experiments. In agriculture, Bessey's own scientific work confirmed his often-stated belief that as botanists, and most scientists for that matter, expanded their scope of investigation they would provide more aid to agriculture. Unlike many other botanists of his day, Bessey believed that botany included cultivated plants as well as wild species, and he chided his fellow botanists for their neglect of cultivated plants. "How much," he queried, "does ordinary botany do in furnishing nomenclature and classification for wheat, oats and Indian corn; for apples, pears, peaches and cherries; for our roses, geraniums and verbenas—in fact for all the plants which have run into many varieties?" He answered that the "student may have in hand popcorn, sweet corn, flint corn, dent corn, and even husk corn, and yet the science of botany gives him but one name—each is ticketed *Zea mays,* of the order *Gramineae.* So it is with every other cultivated plant." Bessey asserted that classical botanists were wrong when they claimed that a "true" botanist was not concerned with plants of the farm and orchard. "If botany is the science of plants," Bessey asked, "how can a plant ever, by any amount of variation, pass beyond its domain? Who shall say that science must stop at this or that line?" Bessey stressed that *"all plants, in all states* and under *all conditions,* should legitimately be included within the domain of botany."[23] What held true for botany, he believed, was the same for the state of most areas of science in regard to what currently was being studied.

As for specific proposals to the regents, Bessey planned a number of both scientific and popular experiments for his faculty. The popular work involved animal breeding and feeding and testing varieties of crops in regard to local conditions. What Bessey classified as scientific research certainly would have been regarded as applied work by many "pure" scientists. Bessey, however, blurred over the popular distinction between pure and applied studies and maintained that if the work was experimental and obtained general results, then it was scientific. He wanted the University of Nebraska to compile weather data and to test the relationship between temperature and humidity upon soil composition, water loss, and seed germination; to chemically analyze soil; and to study the physics, chemistry, and botany of irrigation as well as some of his old favorite topics of injurious fungi, harmful insects, and fertilization in plants.[24]

With the approval of the regents, Bessey and his faculty started the testing and experimentation that were to be associated with the university for the next decade and more. To display some semblance of activity, the Industrial College initially issued five short "press bulletins" during 1885, four of which Bessey wrote on apple blight, a disability in plums, corn smut, and the condition of the college.[25] The next year the dean received money to publish regular-length bulletins, and again he wrote the first bulletin himself concerning the changes being made in the Industrial College, particularly how the charges by the State Board of Agriculture had been corrected regarding the poor condition of the college farm. At best, however, Bessey's research plan for the Industrial College made little progress during its first few years.

Although Bessey successfully had convinced university officials and regents that research was an essential part of the Industrial College, he recognized that adequate funding was necessary. He had long advocated public support for scientific research, explaining to the public and to politicians that such expenditures could return many times the investment by savings from such benefits as prevention of plant and animal diseases. However, he continued to be disappointed in the responses to his pleas.

In 1886, for example, Frank Lamson-Scribner, a scientist for the Division of Botany in the USDA, asked Bessey to help boost the appropriations for the division. Bessey wrote to congressional representatives and persuaded the Nebraska State Horticultural Society to petition the national government to provide for more botanical study within the USDA. The Division of Botany received only $4,500 and the Section of Vegetable Pathology only $3,000 while the Bureau of Animal Industry had an annual budget of $500,000.[26] Botany fared only a little better over the next few years within the USDA, building its budget to $20,000 by 1889, but although this neglect bothered Bessey, he believed that the total scientific budget within the USDA of less than one million dollars was grossly inadequate in proportion to the investment in American agriculture.[27]

Bessey was disturbed not only about the shortage of money, and consequently the small amount of research being conducted in agricultural science in the United States, but also about the type of research being done. In 1886, referring to the "so-called agricultural experiments" being conducted at the nine state and private experiment stations, he stated that "agriculture has been cursed by a greater amount of poor work under the name of experimentation than any other of the great industries." The agricultural work was in such deplorable condition, Bessey further observed, that scientists and scientific journals did not even bother to look at or review publications of the experiment stations because they did not find them worth their while. The one exception to Bessey's condemnation—a place where he believed good work was occurring—was the New York station where E. Lewis Sturtevant was in charge and his ex-student J. C. Arthur was botanist.[28]

To help solve these two major problems of agricultural science—that is, more money and better investigations—in 1882 while still in Iowa, Bessey aided Professor Seaman A. Knapp in writing a congressional bill proposing national experiment stations connected to the state agricultural colleges. Apparently Bessey's contribution was a long list of the type of investigations that the stations were to conduct, specifying both popular and scientific experiments. Certainly Bessey and Knapp had been strong proponents for each type of work, and each was forced to accept the perspective of the other. Representative C. C. Carpenter of Iowa introduced the bill into Congress, but it failed to pass beyond the committee stage. General momentum for some system of national experiment stations was gaining headway, however, and the following year a national agricultural convention organized by Commissioner of Agriculture George B. Loring endorsed the Knapp-Carpenter bill.

As support for national stations increased, much disagreement remained regarding the relationship of the stations to the state, the current colleges of agriculture, state agricultural societies, and the USDA. The debate over experiment stations was part of a general reevaluation not only of what type of investigation was best for agriculture but also where it should be conducted and who should control or monitor it. Knapp had expressed the need for a centralized system that would impose control and research standards over the widely scattered network of stations and college research programs emerging around the country. Like Knapp, Bessey believed a laissez-faire, individualized system would merely continue the type of research being conducted. A number of agricultural colleges, however, opposed such centralization and therefore worked against the Carpenter bill and later similar versions. From a growing variety of bills and plans, an alternative proposal providing more localized control over the stations came out of another one of the national agricultural conventions in 1885. With modifications this became the Hatch Act passed by Congress in 1887.[29]

Although the final act differed in regard to control of the stations, making them more state than national institutions, Bessey generally found the Hatch Act satisfactory except that the $15,000 which went to each state annually from the national government was insufficient for the needs of agricultural science. Some of Bessey's spirit remained in the Hatch Act; the section outlining the research duties of the experiment stations retained his wording.[30]

The Hatch Act became law in March 1887 and Nebraska legislators immediately registered their approval. With $3,500 earmarked by the legislature for agricultural research, and in anticipation of the first $15,000 that the station would receive in March 1888, Bessey presented the regents with another report regarding the additional research the station should now attempt.

He reminded the regents that among his earlier recommendations in 1884 and 1886 few of his original list of four areas of popular and twelve areas of scientific research had yet been started. He therefore proposed that the Hatch money be applied to the areas he had originally outlined. The regents, in accepting the dean's recommendations, named Bessey director of the new experiment station, appointed the six faculty of the Industrial College as the staff of the station, and incorporated the new Patho-Biological Laboratory into the station. In addition, the station was given use of the entire college farm, and as Bessey desired, the station remained a part of the Industrial College.[31]

This last action became a major point of controversy during the early years of the Hatch stations as one faction feared that by leaving the stations attached to the regular scientific departments within the universities the money intended for agriculture would be siphoned off for nonagricultural uses. In Nebraska this thinking complemented the arguments of those persons who wanted to separate the Industrial College from the university. The other side, which included Bessey, wanted the teaching and station staff to be joined in order for each to benefit from the work of the other. Believing that his Industrial College faculty should be researchers as well as teachers, Bessey saw the experiment station as a means of providing money and facilities, as well as justification, for his teachers to engage in research. Separated, it would be more difficult to incorporate research results into instruction. Likewise, in his own situation he believed it essential for the station to have access to the herbarium and library of the botany department rather than to use the limited funds of the station to duplicate these basic references.[32]

Discussing the subject at the first meeting of the Association of American Agricultural Colleges and Experiment Stations in October 1887, those present generally favored keeping the funds of the station separate but otherwise wanted the stations attached to the academic colleges.[33] Most stations were associated in varying degrees with agricultural schools and colleges, but a few were completely separate in administration and staff.[34]

The Nebraska experiment station proceeded slowly. During its first year before the Hatch money was available, the station spent its small legislative appropriation on equipment, but little actual experimentation began. Two bulletins, on irrigation and insects, were prepared during the year but were not published until 1888 when accompanied by two additional studies on animal diseases.[35]

In his own work with the Agricultural Experiment Station of Nebraska, Bessey basically duplicated his duties as botanist for the State Board of Agriculture and continued his collections and observation of fungi, grasses, and forage plants. He published his first report on grasses for the State Board of Agriculture, and in 1889 a bulletin of the experiment station included an article on corn smut by Bessey in addition to articles on smuts and rusts by three of his graduate students. Although serving as director of the station for only two years, Bessey continued as station botanist.[36]

As Bessey continued to provide the foundations for what he believed to be the proper type of department, college, and experiment station, he was caught again in the continuing unsettled administrative situation at the University of Nebraska. While many factors contributed to the situation, one person above all stood at the center of controversy.

In June 1886 a boost to Bessey's plans, and yet a major headache, arrived at the university in the form of Frank S. Billings, a prominent medical veterinarian. The previous December Bessey had recommended to the regents that they establish a course of study in veterinary science. As steps were taken to initiate the course, Billings, who was then residing in the East, approached university officials and congressional representatives from Nebraska and urged them to use their influence to have the Bureau of Animal Industry within the USDA hire him. He wished to be sent to Nebraska to determine the causes of hog cholera and Texas fever that were causing considerable destruction to area livestock. Unable to obtain a position with the USDA, Billings persuaded the Nebraska regents that he was the person who could establish their program in veterinary medicine. He was hired, but not to organize a veterinary school or teach classes. He was to be a full-time researcher and create the Patho-Biological Laboratory, an experiment station for contagious diseases.[37]

Controversy immediately surrounded the work of Billings. On the one hand it brought great praise to the university and the state for its pioneering efforts in combating contagious diseases of animals. Yet on the other hand it was the source of bitter criticism that brought the entire scientific efforts of the Industrial College into question and fueled the critics of the university's agricultural efforts. Billings claimed to have solved the mysteries of a variety of important animal diseases, including hog cholera and Texas fever, but he was unable to provide convincing proof that he was correct. His announced discoveries also ran counter to the findings of the Bureau of Animal Industry, and he soon engaged in a mudslinging public conflict with the chief of the bureau, D. E. Salmon.

Although potentially a good scientist, Billings soon lost the support of most of his fellow workers in the Industrial College, including Bessey, who saw that beyond the veterinarian's personal egotism Billings violated a cardinal principle of scientific procedures by claiming great discoveries before carefully following his preliminary findings with the necessary confirmations. Undoubtedly there was some envy involved. All the faculty in the college wanted more time and money for their own research. Most had little of either, yet Billings did no teaching and was able to corner much of the limited research money provided by the regents. Continuing to push hard for more equipment, assistants, and laboratory rooms, Billings gained control of about one-third of the budget of the new Agricultural Experiment Station of Nebraska after its establishment in 1887. By 1888 the faculty of the Industrial College had largely disassociated themselves from Billings, and University Chancellor Irving J. Manatt and Nebraska Governor John Thayer became open critics of him.

Despite continued support for the Patho-Biological Laboratory by the regents, led by Charles H. Gere and the Nebraska Swine Breeders' Association, pressure against Billings mounted until he left the university in June 1889. By the time he left, politicians, newspapers, and the agricultural societies of the state were embroiled in the dispute, and the uproar provided another opportunity for reopening the charges that the university was misusing money intended for agricultural education. The result was another bill before the legislature to create a separate agricultural college and to dismiss Chancellor Manatt.

Soon extensive lobbying efforts by supporters of the veterinarian resulted in the university rehiring Billings in 1891. In taking the action, the regents committed two-thirds of the experiment station's $15,000 fund to Billings' work; increased his salary above that of his dean; provided an assistant and a chemist; and authorized equipment money that was lavish by experiment station standards.

Billings remained at the University of Nebraska for only two more years, years that proved as turbulent as his earlier stay. The contributions by Billings to veterinary medicine and pathology were extremely paradoxical. Although an active supporter of needed original research in the United States and a proponent of improved professional standards, his constant criticism of the USDA (though much of it was valid and constructive), made the government scientists and agricultural science in general vulnerable to criticism from all sides. While promoting educational reform, original research, and practical scientific application on the one hand, public controversy—on which Billings thrived—on the other hand opened the university to criticism from those persons who believed that any money spent on universities and science was wasted. Billings publicly denounced any element that opposed increased appropriations for the Patho-Biological Laboratory, and his insulting manner often made the defense of scientific expenses by university officials more difficult. In attempting to secure more money for his research, Billings claimed that aside from his own work the university had never accomplished anything of value for the farmers and stock producers of the state. Not only did such statements unjustifiably debase the work of his colleagues, but also again Billings provided fuel to the antiuniversity element within the state and to the persons who disliked what Bessey was attempting to do with the Industrial College.

The regents had removed Manatt as chancellor in July 1888 and appointed Bessey to occupy the position temporarily. Bessey also became dean of the College of Literature, Science and Arts; Lewis Hicks soon replaced him as head of the Industrial College. Bessey remained acting chancellor for three years until June 1891, refusing to take the position permanently.

As a person who was very organized in his thinking and work as well as an energetic and nearly inexhaustible worker, Bessey was able to handle the leadership of the university, but he never admitted to liking administrative duties and tended to view the function of a university from the standpoint of the faculty. He felt uneasy about making decisions that affected the welfare of departments and colleges other than his own, and even as chancellor he often reported needs and problems to the regents but refused to recommend any particular course of action. Effective at formulating and presenting plans for the university, he objected to the necessary lobbying for appropriations with the state legislature. In general, he believed rational persons should be able to follow his reasoning and realize what was behind his proposals and why. He disliked "politicking" from others, and he refused to engage in it himself.[38]

Bessey's strength as chancellor was his ability to bring relative harmony to the faculty, and during his tenure of three years he promoted a number of successful programs and changes within the university. First and most importantly, he moved quickly to persuade the regents to act on a recommendation he had presented in 1888 as dean of the Industrial College: to have all science courses moved from the College of Literature, Science and Arts to the Industrial College. Aside from Bessey's belief that all industrial students should have the same basic science courses as the liberal arts students, his reasoning was that because the Morrill Act required these courses for industrial students, putting all the science students together would avoid duplication of courses and faculty. The immediate practical result of the change was that it gave the Industrial College more than its ten to fifteen students, and the Industrial College could grant the bachelor of science degree in addition to the bachelor of agriculture and bachelor of civil engineering. While the number of agriculture and engineering students did not increase, by adding the former science students from the arts college the Industrial College enrollment jumped to fifty-seven in 1888.[39]

In addition to this fundamental curricular change, Bessey spent much of his time as chancellor trying to quiet criticism of the university from outside. He continued efforts to take more of the agricultural programs of the university to the farmers of the state. Along with Former Governor Robert Furnas he pushed the formalization and enlargement of the Farmers' Institutes through the cooperation of the State Board of Agriculture and the university. Speakers from these two institutions donated their time and the local communities had only to provide expenses. The rapid expansion of the institutes was such that by 1892 university faculty participated in twenty-five such meetings. Next they attempted to secure state funding of the program, and in 1896 the Farmers' Institutes reached an important stage in their development by having a full-time director in the university to organize and operate the program. In addition to his leadership in these efforts, Bessey continued to be a frequent speaker at the institutes.[40]

In other efforts to prove the usefulness of the university, Bessey frequently invited agricultural leaders to the campus for meetings and private discussions.[41] The acting chancellor continued his role as botanist for the State Board of Agriculture and tried to find ways to involve this and other popular organizations in the activities of the university. For example, he had initiated the practice of inviting a committee of members from the State Board of Agriculture and the State Horticultural Society to visit the Agricultural Experiment Station each year to report on its operations.⁴² Although many members of these groups had been supporters of a separate agricultural college and extremely critical of the operations of the agricultural school, Bessey hoped to include them more in its operations. He also hoped to improve the ability of the experiment station to help Nebraskans. To bring the work of the station to more persons throughout the state, he pushed for the establishment of regional substations.⁴³ In general whenever Bessey spoke or wrote throughout the state he was promoting the university and its approach to agricultural science.

In addition to the many speaking engagements with Farmers' Institutes, the Farmers' Alliance, the horticultural society, and a number of livestock organizations, Bessey and other members of the faculty increasingly communicated to the public through the pages of the Lincoln *State Journal* and the *Nebraska Farmer*.⁴⁴ For a short time the faculty even operated the *Nebraska Farmer* with H. H. Wing, professor of agriculture, serving as editor and Bessey and other faculty members holding associate editor positions. Bessey never recorded what duties he had as an associate editor, but he did contribute two series of articles. The first series, "Education for the Farmer's Boy," was an attempt to convince farmers that scientific agriculture was the way of the future and that increasingly it was necessary for farmers to send their sons to the university where such training was possible, and the second was "Diseases of Farm and Garden Crops."⁴⁵

Other achievements by Bessey as chancellor, or at least accomplishments during the time he filled the position, were the securing of appropriations for a new university library, the creation of the College of Law, and the start of a regular summer school. Bessey worked hard for the library, and he helped promote the idea of summer classes by opening his own laboratory to public school teachers for two weeks during the summer of 1889. Other departments followed and soon the summer courses were a regular part of the university and required the appointment of a director, a position which Bessey held for many of his remaining years at the university. Two of Bessey's pet projects that he was unsuccessful in achieving were his proposed quarter system and his idea of moving the campus, with the exception of the law, medicine, and fine arts faculties, to the farm outside the city of Lincoln.⁴⁶

Despite Bessey's efforts to explain the role of the university and to persuade critics that his ideas of a land-grant college were best, the University of Nebraska faced another full-scale attack in the legislature during 1889 when there was another attempt to separate the Industrial

College from the rest of the university. The legislature sent an investigating committee to see that money intended for agriculture was not being misused and to inquire into the inability of the Industrial College to attract students into agriculture. Although Bessey did not come under direct criticism, the program and operations of the Industrial College were thoroughly condemned. The committee objected to the emphasis on scientific training and cultural or liberal arts study, which occupied three-fourths of the student's time, with applied agriculture receiving attention only in the last year. They believed that mixing the industrial and liberal arts students "induced" students away from the farm and studies related to it.

Any attack on a part of the university brought a variety of different forces into the fray, and this investigation was no exception. In addition to those who opposed Bessey's type of agricultural education, there were always persons ready to join because they opposed higher education in general. Also very evident in this investigation were the anti-Billings forces who were quite numerous and extremely vocal.

The committee therefore found a variety of faults in the Industrial College, including misuse of agricultural money and too much money wasted by Billings without verified results. In general they found nothing about the operations of the college that they liked, and the result was another legislative bill to create a separate agricultural college and to sell the university farm. As earlier, the bill failed passage and the status of the agricultural program remained unsettled in the eyes of its opponents.[47] Bessey's only response was to move the scientific courses, in toto, to the Industrial College, which had little to do with most of the objections in the legislative investigation. The Bessey type of agricultural education thus was able to hold the line in Nebraska for at least a few more years until the regents attempted a major reorganization of the agricultural and mechanical programs in the early 1900s.

In 1891 the University of Nebraska finally appointed a new chancellor, James Canfield. In addition to stepping down as chancellor, Bessey asked to be relieved of his deanship of the College of Literature, Science and Arts. For seven years Bessey had devoted a major part of his efforts to administration of the botany department, the industrial and arts colleges, the experiment station, and the university, and his botany had suffered.[48]

Fortunately, Bessey was such an intense worker that his botany had not been totally neglected but only greatly restricted. During the 1880s since coming to Lincoln, he had issued three more editions each of his *Botany for High Schools and Colleges* and *The Essentials of Botany*.[49] In response to a growing concern for botany in the elementary or grammar school, Bessey in 1892 published *Elementary Botanical Exercises,* a small book aimed primarily at teachers. Realizing that most teachers had no training in botany, Bessey provided topics for which teachers could use local plants to introduce students to nature study. Asked by publishers to enlarge the geographical area covered by the book, Bessey resisted changing the scope beyond Nebraska or the prairies. In 1894 he relented and enlarged the book, hoping both to explain the self-training of elementary teachers in nature study and by listing month-by-month activities to provide a guide for the beginning botany course in the high schools. Collecting plants and using a compound microscope—fieldwork and the laboratory—now were combined in introducing students to the general features of the plant world. Bessey added a new second part to the book, a "Manual of the Common Genera of Nebraska." The primary difference between the *Elementary Botanical Exercises* and his other two textbooks was its emphasis upon fieldwork and identification rather than the laboratory. Except for its greater coverage of the "lower plants," his *Exercises* was more like the earlier textbooks of his former mentor and rival Asa Gray, who had died in 1888. In part this shift in emphasis reflected Bessey's drift from physiology and his growing concentration on systematic botany, and in part it was to fill a need for a field guide.[50]

In 1893 a note in the *Botanical Gazette* stated that Holt and Company had "announced from the press" *An Introduction to Systematic Botany* by Bessey. But this announcement proved premature. Nearly every year Bessey reported that the "textbook drags along," and he never brought the systematic study to publication.[51]

Upon coming to Nebraska, Bessey's botanical interests remained channeled basically in the areas that had interested him in Iowa—that is, structural physiology, especially reproduction in "lower plants." Finishing some of the work he started in Iowa, Bessey presented several papers and wrote several short notes in 1885. However, his work in this area, consisting mainly of simple observations, resulted in only some short notes in the *American Naturalist* and increasingly was directed towards issues in classification.[52]

Also of interest in much of this work was the structural relationship between parasitic fungi and their hosts. As before, most of his published efforts in pathology he directed to agriculturists. He started studies on wheat rust and corn smut. While the general nature of these disorders was known, Bessey pursued such issues as what served as an alternative host for wheat rust in the absence of barberry, which was not prevalent on the Great Plains; whether burning straw or alternating wheat fields affected the abundance of rust; the effects of different weather; the extent of rust in newly cultivated lands; whether corn smut injured cattle eating it; the longevity of smut spores; how the smut gained entry into the host; and how best to prevent the distribution of smut spores.[53] In the early 1890s Bessey became interested in another plant disorder that was becoming increasingly pertinent to Nebraskans, the leaf spot of sugar beets.[54] In these efforts Bessey passed along the small but growing knowledge available on plant pathology, but because there was so little known he spent most of his time explaining to practitioners the potential value of pathology and the need for public support to bring these studies to reality.[55] Likewise he continued to stress to other botanists the faults of the old structural botany in which the plant was studied with the intention to identify it rather than to learn how it functioned and hence how parasites affected it.[56]

In addition to pathology, as had been the case in Iowa, upon moving to Nebraska Bessey believed that if he were going to make his knowledge useful to the region, he needed to know the flora of the state. His first step was to start an herbarium of local plants. Botanically Nebraska was of great interest because little was known of the Great Plains. The eastern part of the state was included at least in part in Asa Gray's *Manual,* and some of the western species were described in studies on the Rocky Mountains, but the Great Plains of central and western Nebraska in general were poorly covered.

As the dominant plants of the region and because of their economic importance, grasses and forage plants of the plains most interested Bessey. The 1880s were a time of increased interest regarding grasses and forage plants in general, reflecting a beginning impact of agricultural science on botany. Such writings as Alphonse de Candolle's *Origin of Cultivated Plants* appeared as well as a series of smaller studies on the botany of cultivated plants by the American agriculturalist E. Lewis Sturtevant. Specifically on grasses, George Vasey of the USDA published *A Descriptive Catalogue of the Grasses of the U.S.* and William J. Beal of Michigan Agricultural College provided a popular study on grasses of North America intended for students and farmers.[57]

Immediately upon coming to Nebraska, Bessey started a collection of national and local grasses. With the aid of Herbert J. Webber, a graduate student, he produced an incomplete catalog of Nebraska grasses and forage plants in 1889. At the same time other graduate students in Bessey's charge wrote accounts of the forage situation in specific localities within the state.[58] In the process of organizing this information, Bessey, among others, observed that the grass-covering of the state was undergoing change as settlement and cultivation occurred, and he pushed two projects in relation to these changes. First, he believed that because the native grasses could not be depended upon to support the agricultural needs of the state, cultivated grasses and forage plants had to be found that could withstand the dry conditions of Nebraska. Second, he believed it was possible to cultivate some of the native grasses and forage plants. On both of these points there existed much disagreement among agriculturists, and both of these projects were complex.[59]

Studies were started to determine the value of the native plants, but it was the early 1890s before this information was becoming available. Having determined which of the wild plants seemed best, Bessey encouraged initiating efforts to bring these plants under cultivation. The experiment station started some small plots and used this need as one of the arguments for establishing substations throughout the state. Yet because of finances and the undeveloped experiment station most of this kind of testing, if done, would have to be handled by individual farmers. Therefore, in the popular literature Bessey discussed the steps and difficulties of converting a plant from a wild to a domesticated state and encouraged agriculturists to start efforts in this direction.

The university was able to be more helpful in the other area, that of introducing already domesticated new grasses and forage plants into the state. Again, experimental plots were grown on the station grounds, but what was successful in Lincoln had limited applicability for the rest of the state. Particularly the drylands presented the major problem. Once more Bessey had to rely on the experiences of farmers and ranchers as the basis for most of his recommendations in this area.[60]

During the 1890s Bessey tailored his applied work along the same lines. Topics concerning native and cultivated grasses, forage plants, and wood resources resulted naturally from work with the new Botanical Survey of Nebraska and were a continuation of his attempts to collect and identify the state's beneficial and harmful plants, to test plantings in various conditions and sections of the state, and to determine the food value of the various regional grasses (see Chapter 6). His concern here was with such problems as the disappearance of wild grasses, the types of hay suited to dry farming, the danger of imbalance between cattle and forage plants, and the danger of overgrazing the Sandhills of Nebraska. Each year in his reports he stressed the slowness with which this collection of information was occurring and promised an eventual illustrated and comprehensive bulletin on these topics.[61]

As the decade progressed, Bessey voiced strong warnings about the depletion of the state's grasslands and the threatened loss of such a valuable resource. The wild grasses and pastures were vanishing faster than replacement grasses could be found, and he observed increases in overgrazing and overproduction of crops. Aside from losing the grasses as a food source for animals, the loss of roots threatened the soil of the windy plains. In 1897 Bessey visited western Nebraska, a region many farmers and ranchers were abandoning at that time due to drought and economic depression, and saw the vast damage that farming was doing to such a fragile area. He resolved to demonstrate that with proper care and practices much of the state's grasslands, although unsuited for typical crop farming and grazing, could be restored and that irrigation could be increased.[62] Although many persons had doubted that most currently known cultivated forages would grow successfully in Nebraska, by the 1890s bluegrass, timothy, millet, alfalfa, and clover were replacing the native grasses. Aside from popularizing these plants in the agricultural press, Bessey provided little else in the way of direct help.

Another problem within the state during the 1890s that required Bessey's attention was weeds. Particularly of concern to farmers was the rapid spread of Russian thistles. Bessey responded to the many inquiries by suggesting to the governor the need for a convention and for legislation to combat the new invader. Bessey lectured at farm meetings, displayed the plant at the state fair, and issued an experiment station bulletin informing farmers about how to identify the plant, the means of its distribution, its life cycle, and some methods of combating its growth. The bulletin also provided a copy and discussion of the Wisconsin weed law that Bessey proposed as a model for Nebraska to follow. These actions were successful; the panic over the Russian thistle subsided after three to four years, and Bessey reported that it was no longer a danger to the state.[63]

In other applied work, Bessey and particularly some of his students took an interest in such topics as fertilization, hybridization, and transpiration in regard to the state's plants. On the latter topic, as Bessey's attention turned more and more to the semiarid region of western Nebraska, he pointed to the need for research on problems unique to the agriculture of that region, such as the mechanism of water loss in growing crop plants and the effect of irrigation on the structure of the plants, especially their roots.[64]

In line with his continuing interest in structural physiology, Bessey during the 1890s wrote a series of popular articles mainly for the State Horticultural Society. The articles covered botanical aspects of the apple, grape, strawberry, plum, and cherry, the structure of wheat grain and bran, and honey-producing plants. Drawing from his own observations, but mainly from the work of his graduate students and from other agricultural botanists, he illustrated a variety of ways in which plant physiology affected horticultural practices. To grow the best fruit possible, Bessey exhorted growers to know more than traditional practices and trial-and-error methods. The horticulturist needed to understand such information as how soil and temperature affect growing, how plants obtain food and grow, how the flower functions, how disease affects plants, the process of pollination and fertilization, and the process of water storage and water loss in plants.[65]

A continuation of Bessey's earlier agricultural interests was his attempt to keep agriculturists informed about new research in plant pathology. Bessey noted that Nebraskans were fortunate in that plant diseases as yet "rarely do enough harm to warrant any great outlay of time or money." He did add, however, that when there was an outbreak of any kind he did not have sufficient time or money to seek causes and solutions with the one hundred dollars allotted to botany by the experiment station.[66] In an attempt to reduce the burden of having to answer numerous letters regarding diseases, he provided a general description of plant ailments in the report of the State Board of Agriculture, and he followed over the next few years with short descriptions of specific diseases in Nebraska.[67]

To overcome the lack of time and money for agricultural botany, Bessey in 1897 requested an assistant botanist for the experiment station whose time could be divided between research for the station and teaching in the Industrial College. He explained to the administration that while a consulting botanist from the university, as he had been, had generally been satisfactory up to this time, he now believed the station needed a full-time researcher as more botanical problems became pressing. In particular he pointed to the introduction of irrigation and to the lack of botanical knowledge in relation to this. Also at this time, Bessey finally was able to hire a stenographer to help handle the extensive correspondence related to these practical questions.[68]

In Nebraska there continued to be concern with losses from corn stalk disease and, as the sugar beet industry increased, from leaf spot. Fungal diseases had always interested Bessey. Unlike his earlier colleague, Frank Billings, Bessey did not believe that the same bacteria involved in plant diseases could cause disease in animals, a point that was at issue in corn stalk disease and similar fungal growths on forage plants.[69] In particular, Bessey investigated physiological aspects of rusts, which were of growing scientific and economic interest in the 1890s.[70] In addition to making his own observations, he cooperated with other scientists and with the USDA to gather information on the national distribution of rusts. In 1898, for example, the botany department at

Nebraska assisted Mark A. Carlton of the Division of Vegetable Physiology and Pathology of the USDA study rusts in Nebraska.[71]

Throughout the 1890s there also was increased pressure on the experiment station to provide information on poisonous plants, a topic closely related to diseases from fungi. Bessey noted that by the end of the decade "every year so many inquiries come to the Experiment Station regarding what plants are poisonous that it has become necessary to take up the matter somewhat more critically and exhaustively." Typical of his method of gathering such information, through the pages of the *Nebraska Farmer* he asked agriculturists to send specimens of all suspected poisonous plants.[72]

By the turn of the century, Bessey was able to turn over most of the work connected with the experiment station to other persons, but for a few more years he had only limited success in relieving himself of many of the other agricultural duties connected with his work. However, by using his writings for two or three related purposes he was able to issue bulletins for the experiment station, reports for the State Board of Agriculture, addresses to Farmers' Institutes, short notes for agricultural newspapers and journals, and papers for the State Horticultural Society. It was understandable that there was much duplication in these writings and that most of the information was culled from literature and correspondents rather than from his own investigations. These writings were of an informative nature, popularly written, and meant to aid agricultural practitioners. Yet they provided the farmers, fruit growers, and ranchers who bothered to read journals and bulletins with up-to-date summaries of research that was as yet unfamiliar to many professional botanists in America.

A devotion to the active dissemination of scientific knowledge enhanced Bessey's contribution to agriculture. From his first year of college teaching at Ames onward, Bessey had a strong sense of obligation to make scientific information available to agriculturists. The college classroom, while important in this process, was not sufficient. Thus, until late in his career when the university had grown to the size that he had assistants and other professors to replace him, Bessey personally continued to provide such services.[73]

Bessey was a nuclear figure in a network that included the USDA, university and experiment station researchers, his own students, members of the state agricultural and horticultural societies, and practitioners. Through his public speaking, teaching, and writing Bessey channeled the growing body of information back and forth throughout this network. All parts of the network provided him with information, and in turn all parts benefited from their association with him. He relied on the work of the new professionals as found in the reports and bulletins of the USDA and the state experiment stations, the agricultural press, and the proceedings of the Society for the Promotion of Agricultural Science.

Bessey's connection with the USDA became even closer in the 1890s as increasing numbers of his graduates acquired positions in the Bureau of Plant Industry. Albert Woods joined the Division of Vegetable Physiology and Pathology in 1894 and later became chief of the division, Jared Smith was hired by the newly formed Division of Agrostology in 1895, and more soon followed.[74] Bessey observed, "Isn't it funny how botany boys of the University of Nebraska turn to agricultural phases of it [botany]. I delight in it, and think it is just what ought to be."[75] Actually, by the 1890s Bessey was training his botanical students with national service in mind. He believed that few better places existed than the USDA where a beginner could find as good an academic and research environment. In turn, the Department of Agriculture found Bessey's students well suited for its needs. His students received a broad botanical training with an emphasis in physiology and pathology, were strong in laboratory techniques, and had a good foundation in bibliography. As a result of this preparation, Bessey's students could adjust to research problems whether they involved cotton diseases in Texas, diversified farming in Hawaii, lumbering in Pennsylvania, or rice diseases in South Carolina.[76]

In his relations with the USDA, Bessey followed the federal department's work closely and took an active role in formulating ways for the USDA and the agricultural colleges to benefit one another. During the late 1880s, Bessey provided materials for an exhibit by the USDA on the rate of tree growth, collected mushrooms, offered the services of the Patho-Biological Laboratory, collected information on the distribution of the English sparrow in Nebraska, provided specimens and information on crop fungi, and frequently collected plant specimens desired by the national institution.[77] Bessey was pleased to see the USDA increase its attention to grasses with the establishment of a separate division for them in 1893. Other concerns in which Bessey cooperated with the USDA included a near-crusade to end the free distribution of seeds by members of congress which Bessey and others recognized as a political rather than an agricultural endeavor; his help with the study of rusts; and his arranging for the trial of Russian seeds in various parts of the state.[78]

In large part due to the efforts of Bessey, enrollments in the Industrial College at the University of Nebraska increased during the nineties. By the end of the decade the number in the college approached five hundred students, and Bessey immodestly informed the public that they maintained "this large enrollment in spite of the fact that our conditions for admission are nearly two years in advance of those in other colleges in the country." Still, relatively few of this increased number were agricultural students until after the turn of the century.[79]

Bessey continued as a leader in the effort to resist the separation of the agricultural program from the remainder of the university. To Bessey, respectability for agricultural and industrial education depended upon not having separate classes and colleges for these programs. His efforts also reflected his view that in American education "the longer I teach the more I am impressed with the value of training all classes of people together."[80]

In reaction to continued criticism that the Nebraska agricultural program—with its three years of cultural and basic science courses and only one year of applied courses—lured students away from the farm, Bessey answered that the role of the university was not simply to train students and return them to the farm, but to give rural students the same kind of broad, cultural education that other students received and to provide a foundation of science that a student could then take to any desired occupation. In his own botany classes, which by the late 1890s numbered around 120 students in seven different courses, Bessey estimated that probably less than one-fourth intended to pursue horticultural or agricultural fields.[81] Bessey emphasized it was not the fault of the university curriculum that many rural students chose not to return to farming, pointing out that "the University does not change these young men, they do themselves . . . you can't compel American boys to follow any particular line of work." He added that "we have tried honestly to carry out the spirit as well as the letter of the law and it pains us to be charged with dishonesty."[82] The university countered many of the objections to its agricultural program by creating a new School of Agriculture in 1895. It was a secondary school that emphasized a practical course of study intended to return its students to the farm. Still, other than this, agricultural studies at the University of Nebraska continued to struggle.[83]

# 4
# Seeking Maturity: The "New Botany"

IT WAS DURING the two decades of the eighties and nineties that American botany largely became what it was to be for the next half-century in terms of subject orientation and professional institutions. The "new botany" movement embodied a variety of desired changes and helped achieve an important level of maturity in American botany. During the 1870s and 1880s American botany grew from a small group of largely amateurish plant collectors and identifiers into one of the most energetic and expanding groups of American scientists by the 1890s. In large part this growth and energy were due to increased opportunities for scientific occupations in the emerging land-grant colleges, agricultural experiment stations, and the scientific activities of the USDA. Also, the attention generated by the evolution issue as well as the increased impact of German botany and education on Americans helped botany attract a larger number of capable and professional-minded persons. Emerging to advance and to embellish upon these general developments was a group of dedicated young botanists who envisioned making their discipline respectable, attractive, and even dynamic. They desired both professional growth and a change in the emphasis of botany from taxonomy to a new morphology, physiology, and pathology.[1]

Particularly during the early 1880s, the "new" botanists encouraged changing the subject matter of American botany. Whenever possible proponents such as J. C. Arthur, William J. Beal, John M. Coulter, William Farlow, Joseph Rothrock, William Trelease, Lucien Underwood, and, of course, Charles E. Bessey advocated teaching and studying botany as a science of living plants instead of the herbarium specimens of the "old" taxonomist. As Coulter viewed it, "One must discover in a general way how the individual plant lives, for the plant covering of the earth's surface is a living one, and plants must always be thought of as living and at work."² These proponents wanted botany to be an experimental and laboratory science; German botanical laboratories in particular impressed visiting Americans with their stress on anatomical and physiological botany. Although morphology and physiology were not new to botany, the young American botanists believed that "new" investigations should be more comprehensive than before and "different" from what previously was studied. Morphology should now stress more internal structure—the cell, organs, and tissues—and the stress also should be more comparative with concern for tracing descent. Thus, botany was no longer to be a "pleasurable pastime of identifying the plants of the neighborhood" or merely the placing of dead plants in presses and on herbarium paper; botanical advancement must now be based on the "experimental method of inquiry."³

Although the new botanists wanted an expanded view of botany that would center around morphology, physiology, pathology, and plant geography, most did not want to abandon systematic botany altogether. They did, however, see that systematic studies limited the perspective of Americans to the wider scope of the discipline and that systematic botany as practiced by many Americans would never result in anything more than artificial systems of classification. Joseph Rothrock, a botanist at the University of Pennsylvania, observed, "Systematic botany must, if it represents a strictly natural system, be founded on a nice appreciation of the entire organization, the life history of the individual, and its relation in present and past time to allied plants. This then is the highest, all embracing trend botanical thought can assume."⁴ New botanists hoped, therefore, to change the nature of taxonomic botany by making it an evolutionary study of the origin and relationships of species.

In all this change of emphasis in the subject matter of botany, although the German influence was acknowledged and probably the most influential, to many Americans the English naturalist Charles Darwin was seen as the inspiration as well as the symbol for the changes the new botanists desired. As Bessey pointed out, "Darwin led and where he did not enter himself he pointed out the way." Although evolution as explained by Darwin was certainly an integral part of the "new botany," equally significant as a model for the type of studies needed were Darwin's experiments that provided his underlying evidence in *On the Origin of Species* and upon which he expanded in his later work. Particularly important for botanists were his publications *Variation of Animals and Plants under Domestication, Insectivorous Plants, Power of Movement in Plants, Different Forms of Flowers on Plants of the Same Species, Climbing Plants, Fertilization of Orchids,* and *Effects of Cross and Self Fertilization in the Vegetable Kingdom.* Asa Gray captured the direction of this type of botany in his children's book, *How Plants Behave: How They Move, Climb, Employ Insects to Work for Them, etc.* Botany as the study of living organisms would then differ little from zoology in what it studied: physiology, the life processes involved in obtaining food, assimilation, growth, and reproduction; morphology, the structures (leaf arrangement, vascular systems, roots, reproductive organs) that allowed the organism to carry out its life processes; and geography, the environment in which the organism operated. The new botanists saw in this approach the basis for an experimental science, placing botany alongside the other sciences, and the basis of dynamic discoveries that would replace the more mundane collection and description of plants. As Darwin had indicated, the new botanist needed to view plants as biological problems rather than as something to catalog.[5]

Professionalization of botany went hand in hand with the transformation in subject matter, and the improvement and development of new institutions were an integral part of the "new botany." In the mid and late nineteenth century many disciplines in the United States, both within and outside science, developed greater self-awareness. The advocates of professionalism desired to establish credentials and academic programs to train specialists, found new societies, promote journals, and appeal for public support of their work.

Bessey became actively involved in all these areas of change during the 1880s, and the basis for his role was established primarily with the publication of his first textbook and his subsequent editorship of the botany section of the *American Naturalist.* The new botanists believed that journals were certainly a key to bringing about the desired changes in their discipline. During the 1880s foreign journals remained essential, but the American ones generally improved. The *Bulletin of the Torrey Club* still covered only systematic botany, but it expanded the range of its interest beyond the local flora of New York City to a somewhat more national coverage. The *Botanical Gazette,* edited initially by John M. Coulter, who later was joined by J. C. Arthur and Charles Barnes, was envisioned by its editors as the journal of the "new botany." Coulter was a professor of natural history at Hanover College when he started the journal in 1875 but moved to Wabash College in 1879. Following short terms as president of Indiana University, 1891–93, and Lake Forest College, 1893–96, he headed the botany program at the new University of Chicago. Barnes, an undergraduate student of Coulter's at Hanover, taught at Purdue and the University of Wisconsin before joining Coulter at Chicago in 1898. Arthur, one of Bessey's undergraduates in Iowa, replaced Barnes as botanist at Purdue.

Coulter had indicated to Bessey in 1880 that he hoped to make the *Botanical Gazette* entirely a physiological journal, but he could not get botanists such as Bessey to write for it. "You are to be congratulated upon your success in giving us a lift from the mire," Coulter noted in reference to Bessey's first textbook. "Can you not follow this up by more physiological work in the shape of notes to the *Gazette*?" In looking to Bessey, among others, Coulter pleaded that the "young botanists of the country should have their thoughts turned into other channels than finding new stations [locations], pressing and drying and exchanging, with the dim hope ever before them of finding something to which their name can be attached. . . . I would like to make an onslaught on all this old fogy work."[6]

In 1881 Bessey complied with Coulter's request by submitting two short items on physiology to the *Botanical Gazette*. In the first Bessey described several machines he had built, similar to one developed by Sachs, for measuring the longitudinal growth of plants. In the second he summarized the findings in a thesis by one of his graduate students, and Coulter added a third note on physiological work done by students in Bessey's laboratory.[7] Primarily, however, Bessey answered Coulter's challenge to promote the "new botany" not through the *Botanical Gazette* but through the botanical section of the *American Naturalist*.

Zoologist Alpheus S. Packard, Jr., of Massachusetts Agricultural College and later Brown University, and Philadelphia paleontologist Edward Drinker Cope had acquired the *American Naturalist* in 1878, and two years later, in the same letter in which he complimented Bessey on his new textbook, Packard asked the Iowan to take charge of the botany department of the journal. In accepting Bessey realized that a revitalized journal such as the *American Naturalist* could have significant influence in promoting the type of research and teaching that he believed was a necessary foundation for American botany. Also, his own reputation certainly would be helped by having his name constantly before the readers, and although there was no salary, compensation for his work would come in the form of other botanical journals that were exchanged with the *Naturalist*.[8]

A monthly publication, the *American Naturalist* contained articles, departmental sections for each major division of science, news of societies, abstracts of articles from other journals, scientific news, and editorials. Bessey's duties were to screen all manuscripts on botany, to arrange for or write reviews, and to write or compile material for short items on botany entitled "General Notes." Finally, he wrote the "Botanical Notes," a potpourri of information such as interesting or unusual observations, summaries of research, and brief comments on current literature. The coverage in the botanical section displayed the efforts made by Bessey to keep abreast of current literature, and he kept his readers informed of the broad sweep of botany. Reporting not only on American scientific journals, he also kept pace with the varied publications of the United States government, the land-grant colleges, and many journals in England, France, Germany, and Italy. During his editorship Bessey usually reviewed three to five books a year, and although most of the research notes were the work of other persons, he often commented on the findings. The notes he wrote himself covered a wide variety of topics, but as would be expected, he concentrated on topics concerning fungi, classification, experiments in physiology, and suggestions for laboratory teaching.

In addition to his editing duties, Bessey wrote two articles for the *Naturalist* during his first year of association. Both were summaries of botanical writings by Americans during 1879 and 1880. Such summaries of "progress" were fairly common features of the journal. The summaries reflected the general state of American botany although Bessey confined most of his evaluations to the individual writings. As for the numbers of articles cited, physiology and the "lower plants" compared favorably. There were nearly three times the number of citations for algae, fungi, mosses, and ferns as there were for flowering plants, yet in terms of length more space was given to flowering plants. Of course numbers and space alone were no indication of worth, and Bessey did not comment on the relative value of the work done in these areas. More revealing, regarding the topical divisions within botany, was the fact that for every five or six articles in systematic botany there were only two in structural physiology and one in plant geography. The dominance of systematic studies was further demonstrated by Bessey's evaluation that even though a significant number of articles appeared in physiology, these were mostly based on a "few quickly-made observations" and included little on "micro-anatomy and proper physiology." Bessey excused this caliber of work, however, by judging his own situation at Iowa Agricultural College as typical of the overload of nonbotanical duties that most American botanists had to endure.[9]

Through the 1880s Bessey maintained a relatively large botanical section, writing an average of one or two of the items himself each month. While keeping a helpful and informative botanical section throughout most of this period, the journal itself had troubles from the mid-1880s onward. First the coeditors had a parting of the ways, with

Packard leaving, then the publication faced unusually severe financial problems. By 1890 the journal again was operating with reasonable success, but in 1891 the botany section dropped off sharply when Bessey assumed the position of acting chancellor of the University of Nebraska. One-third of the issues for 1891 had no botany section, and similarly there was no section for two or three issues each year from 1892 through 1894.[10]

Initially the greatest praise for Bessey's handling of the botany section in the *Naturalist* came from editorial comments by John M. Coulter in the *Botanical Gazette*. Having complimented Cope and Packard upon their selection of Bessey, Coulter noted that "Bessey is making the Botany Department of the *American Naturalist* more valuable than it has been for years. It is kept abreast with the times and botanists get hints of all that is doing in the botanical world." Coulter also copied some of the "Notes" by Bessey for the *Gazette*. At the end of 1882 Coulter added that "the *American Naturalist* closes its 16th volume and is well entitled to the position of the most popular scientific periodical in the country. The department of botany, under the direction of Professor Bessey, has been a great success during the past year." Coulter also kept his readers informed on Bessey's various activities such as his summer course at the University of Minnesota in 1881, the contents of his laboratory course, his vacation in the East, the writing of his new textbook, the type of physiological and laboratory work done by his students, and the nature of his latest research and writing. In turn, Bessey promoted the *Botanical Gazette* and praised Coulter for his efforts to advance the "new botany." What was developing was an informal partnership, a professional friendship, that over the next three decades was to have profound influence on American botany.[11]

Coulter and Bessey complemented one another in many of their interests and ideas, and with Bessey in charge of the botany of the leading national journal and Coulter heading one of only two American publications devoted solely to botany, they were in strategic positions to promote the "new botany." In the *American Naturalist,* Bessey challenged botanists to raise their profession to the level of other American sciences; then they could move to be respectable in relation to European botany. In particular, in the early 1880s Bessey and some of the other promoters thought that American botany needed an institutional home, and most recommended the American Association for the Advancement of Science (AAAS) as the best place.

Coulter, Arthur, and Barnes joined Bessey in promoting the AAAS as the professional home for botanists through editorials in their respective journals. Each year they recorded the progress botanists made toward their view of respectability, and they used the showing at the AAAS as the major basis for their judgment. Prior to the early 1880s botanists were sparse in attendance at the AAAS and their papers made little impact. Anticipating the 1882 meeting, for example, Coulter scolded the botanists: "Zoologists, geologists, chemists and mathematicians flock to it [AAAS] in great numbers, but botanists are both few in number and modest in spirit."[12] Coulter confessed that this was to be his first AAAS meeting. Bessey, who had been a member of the AAAS since 1872 and a fellow since 1880, apparently attended only one meeting during the 1870s, but like Coulter he became active during the eighties. He presented his first AAAS paper in 1882 in Montreal and followed with a second paper in 1884 in Philadelphia.[13]

In their journals Bessey and Coulter devoted particular attention to the AAAS meetings in 1883. The coming meeting in Minneapolis was the first in the "West" since 1872 and offered midwesterners an opportunity to demonstrate their growing numbers and to "prove their worth" in the scientific community. Bessey and Coulter were not disappointed with the results. Botanists had an all-time high attendance and for the first time botany papers outnumbered those on zoology in the Biology Section.

At that same meeting, thirty persons decided that botanists needed an organization in addition to the paper sessions, which they shared with the zoologists. Consequently they formed the Botanical Club. Following the format and example of the Entomological Club of the AAAS, the Botanical Club was to be informal, having no constitution although it required membership in the AAAS. Beyond the social purposes and the planning of field trips, the intent of the club was unclear. Some proposed having meetings of the club in which members would read and discuss additional papers; these meetings would be held when they did not conflict with the Biology Section. Opposing this suggestion were those persons who did not want botanists to become too specialized or secluded from other biologists and who feared that papers read to the club might distract from the importance of the Biology Section. It did seem clear, however, that founders of the Botanical Club intended it to include a function of conducting some business that affected only botanists. In this matter they hoped to use the club to promote botany and to pursue problems facing such growth. The founders moved into this function during their initial meeting by appointing a committee consisting of Bessey, Coulter, and William Farlow to investigate and pursue the possibility of getting special postal rates from the government and obtaining adjustments on postal regulations for shipping botanical specimens.[14]

In offering explanations for the good showing of botany at the Minneapolis meetings, J. C. Arthur acknowledged the field's professional growth and particularly the importance of the journals in promoting the work of the AAAS. In addition he believed there was a growing sense of community among the botanists, who were beginning to want to meet one another and to exchange ideas. Finally, Arthur contended that the success of the meeting was "directly traceable" to the Summer School of Science organized two years previously at the University of Minnesota. A "distinctive feature" of the botanical part of the summer school, which Bessey had organized and taught, was its study of "all grades of plants" with its laboratory that, in drawing on the local region for its materials, Arthur believed had stimulated the curiosity and excitement displayed by local botanists at the AAAS meeting.[15]

Not everything about the 1883 Minneapolis meeting pleased its promoters, however. Coulter was still bothered by the caliber of the botanical papers. Too long and too careless in their preparation, he concluded, "some were mere essays about well known facts, and most were observations about such trivial things that they could hardly be called profitable."[16] Coulter's criticism revealed a growing problem regarding the nature of the AAAS. One view of the AAAS was that the meetings should help popularize science, and the papers should therefore be understandable to the large number of interested amateurs who it was hoped would come to the meetings. The proponents of professionalism posed an opposing view that the AAAS should increasingly reflect the specialization necessary for respectability in the eyes of European scientists. Indecisive as to which view the botanists should follow at this time, Coulter stressed the importance of popular support and the continued role played by amateurs, but regardless of which path the AAAS followed, he was emphatic that the papers presented, whether technical or popular, be reports of original research.[17]

Bessey was likewise very active at the 1884 meeting of the AAAS in Philadelphia. He served as secretary of the Biology Section and was elected president of the Botanical Club in addition to reading two papers. Reporting for the postal committee appointed by the club the previous year, he conveyed the committee's failure to convince the Postmaster General to change the rates on botanical specimens. Finally he was appointed to a new committee of the Biology Section—along with Farlow, Arthur, William Beal, Joseph Rothrock, Thomas J. Burrill, and Charles H. Peck—to encourage more research on plant diseases.[18]

Again, both Bessey and Coulter reported dissatisfaction with the caliber of papers delivered by botanists. Bessey, observing that botanical papers at the AAAS had "far less depth" than zoological ones, pondered, "Have not the botanists for so long thought of discovery as connected with new species that they have forgotten that there may be room for brilliant discovery in structural and physiological fields also?" Bessey likened the type of simple systematic studies in botany to the study of skins in zoology.[19] Despite some signs of vitality among American botanists, much of the apparent change to the "new botany" was often more a promise of better things for the future, and Bessey and Coulter were impatient with the slow rate of change taking place by the mid-1880s.[20]

Bessey observed optimistically in 1885 that "one of the hopeful signs of the times, so far as botany is concerned, is the increasing interest taken in the study of the lower plants in this country."[21] Bessey noted that while the desire of "progressive" botany was to research all forms of plants, even systematic study was difficult to conduct in the United States. Most groups of plants lacked basic, comprehensive descriptive works, and where some studies existed they generally covered only a part of the country. For many plant groups, foreign systematic manuals had to suffice.

On fungi, for example, there were no general American manuals, and the few local or family studies were so scattered through a variety of journals and miscellaneous public and private publications that few persons could keep abreast of them, let alone afford to collect them all. As a result, American mycologists had to rely primarily on M. C. Cooke's *Hand-Book of British Fungi,* which according to Bessey had "imperfect" descriptions and "antiquated" classification. There was not even a complete manual on American flowering plants; the various manuals by Asa Gray, A. W. Chapman, Sereno Watson, and John M. Coulter were regional in coverage. It was generally understandable to the promoters of the "new botany," therefore, that progress would be slow when so much fundamental information was lacking.[22]

Among other problems hindering the study of "lower plants" was that few persons were being trained to replace the few existing specialists when they died. According to Bessey, this resulted from the training botanists received in the United States, which still emphasized flowering plants.[23] However, there were some positive indications as far as teaching and the training of botanists were concerned. For example, Bessey liked what he saw in Gray's revised edition of his much-used *Lessons in Botany.* Gray changed the title to *Elements of Botany,* which had been the title of his first textbook fifty years earlier, and in a review Bessey compared the two books. Bessey noted that "there is a good deal of similarity between this pioneer and the book which now, after the lapse of half a century, bears its name; and still there are very many differences." A major difference, Bessey observed, was that "vegetable physiology was very crudely treated in the earlier book," and there was

c. 1880     November 1899     c. 1905

# CHARLES E. BESSEY

1914

*Sem Bot was an informal secret society formed and operated by advanced botany students of the University of Nebraska in the 1880s. Under the guidance of Bessey, it developed into an honorary society.*

Sem Bot, c. 1890

Sem Bot, 1896. Bessey is seated second from the right.

Sem Bot, February 1907. Bessey is in the front row, third from the right.

Minnesota Seaside Marine Station, Vancouver, c. 1895. Bessey is standing second from the right.

As a pioneer of the "new botany" in the United States, Bessey stressed the study of living plants. The university greenhouse provided a place of study and supplied plant specimens for the laboratories.

Physiological Botany Laboratory, April 1914. The laboratory was the center of instruction in Bessey's department. Bessey designed much of his own furniture and equipment and carefully planned the arrangement of his laboratories.

# Scenes from Bessey's botany department at the University of Nebraska

Beginning students in Botany 1 and 2 at work in the General Botany Laboratory, April 1914. Bessey used the laboratory for both beginning and advanced instruction.

Botany staff, 1913–1914. Bessey, Head Professor, is fourth from the right in the front row. The faculty included Raymond J. Pool, Associate Professor (front row, fourth from the left); Margaret Hannah, Instructor (front row, second from the right); Leva B. Walker, Assistant Professor (back row, third from the left); Elda R. Walker, Associate Professor (back row, third from the right). Other departmental staff included fellows, scholars, a collector, an assistant clerk, and a storekeeper.

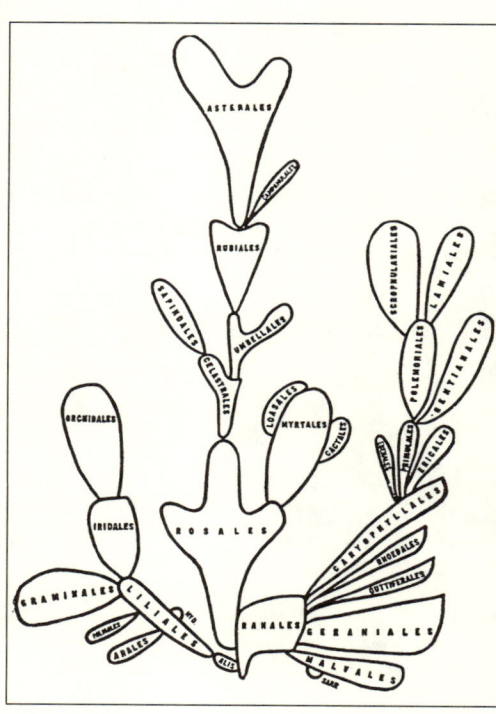

[ABOVE] While known most for his use of the laboratory, Bessey never lessened the commitment of his department to fieldwork. Established and directed by Frederic and Edith Clements, the Alpine Research Station at Manitou Springs, Colorado, was a place for study and summer fun for University of Nebraska students.

[LEFT] Bessey's chart to illustrate the relationship of the orders of flowering plants. This chart has the characteristic appearance of a cactus, which earned the nickname of "Bessey's Cactus." [From Charles E. Bessey, "The Phylogenetic Taxonomy of Flowering Plants," *Annals of the Missouri Botanical Garden* 2 (1915): 118.]

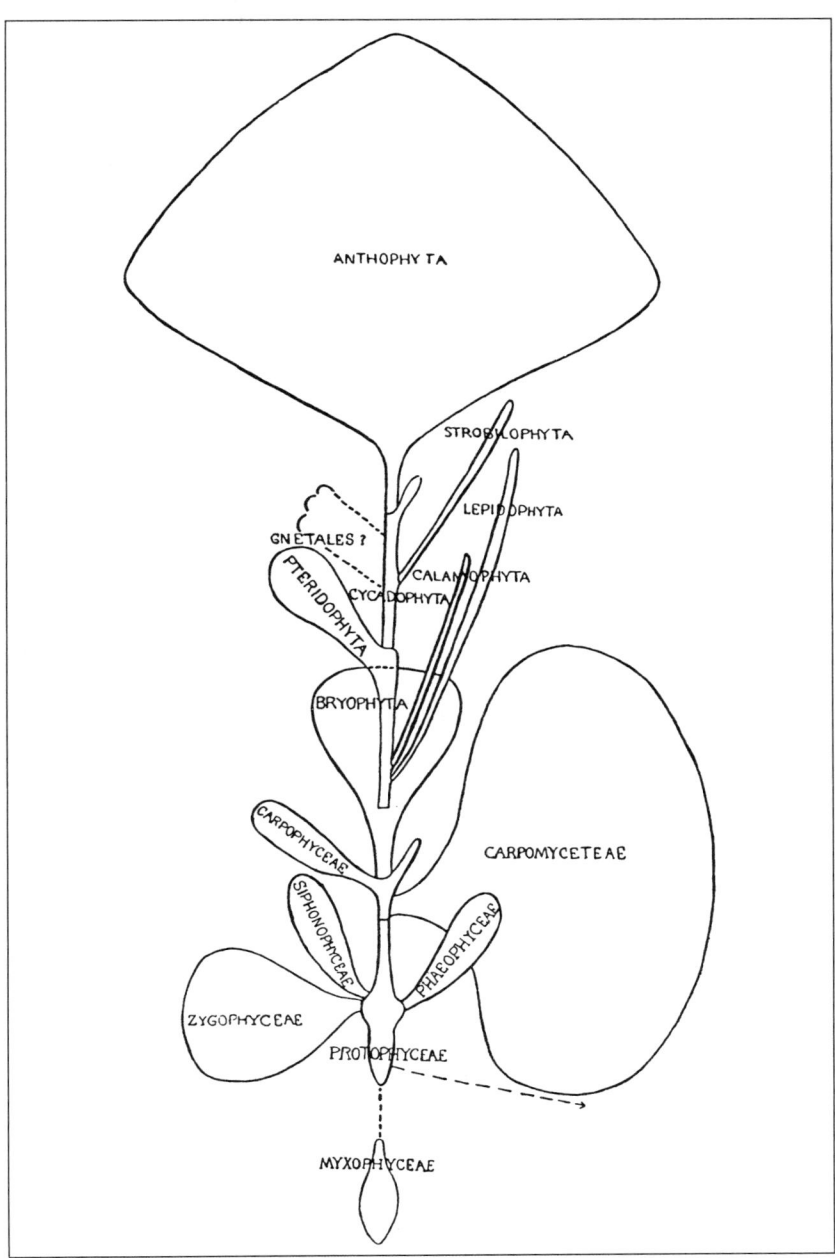

Bessey's chart to illustrate the relationship of the major plant groups. Bessey devised his first chart in 1893 to help students visualize the relationships. From his first straight-line drawings, Bessey improved his charts by using jointed segments showing the points of deviation of one group from another. The size of segment was determined by the number of families or orders in each and its shape by the amount and direction of deviation. [From Charles E. Bessey, "A Synopsis of Plant Phyla," *University Studies* (University of Nebraska) 7 (1907), plate I.]

Charles E. Bessey in his office, c. 1912

nothing about the content of cells. In contrast, "in the present work protoplasm, cells, cell-contents and cell-walls, receive sufficient attention to give the beginner a general knowledge of what they are." Bessey continued, "A great advance" was that "which we may call Darwinian botany." Also, the changes in the coverage of the nonflowering plants was "striking." On this last point, a *Botanical Gazette* reviewer whose opinion was very similar to Bessey's emphasized that while improved in its general coverage of the entire plant kingdom, the book still needed to give more attention to the "lower plants" to achieve proper balance.[24]

Also on the positive side, the late 1880s witnessed the start of the *Journal of Mycology* devoted to American fungi, the publication of several improved and comprehensive works from Europe that provided a much-needed general systematic outline of the fungi, and the start of "The Host-Index of the Fungi of the U.S." by William Farlow and A. B. Seymour.[25] In 1885 and 1886 botanists attending the AAAS registered with the Botanical Club by their special interest, and the flowering-plant people outnumbered the nonflowering by only thirty-six to twenty-nine.[26] By 1891 only about two articles on flowering plants were published in the *Botanical Gazette* for every one on lower forms. Despite these improved conditions, most of Bessey's comments still concentrated on the needs and shortcomings of American work in this area.

The new botanists saw other hopeful signs by the late 1880s. There was more available in terms of physiological studies for the general writers and teachers to learn from. In addition to Bessey's large textbook and the English edition of the German text by Julius von Sachs, Americans during the course of the 1880s had access to translations of Karl Prantl's and Anton de Bary's German works on physiological botany, and in 1885 and 1886 the American botanist George Goodale of Harvard produced two volumes on structural physiology. Bessey was pleased that Goodale's work represented the results of German methodology and provided Americans with a good summary of the recent research from European laboratories.[27]

The publication by Goodale stirred the editors of the *Gazette* to announce that "botany in America was never in a more flourishing condition." They maintained that while American work in systematic botany had been in the "front rank," the work in other areas had seriously lagged. They continued that "the study of anatomy, development, and the habits of plants received great impulse by the advent of Sachs' Textbook in 1875 and it was especially promoted by Bessey's *Botany* in 1880 . . . the latest addition to that line of texts, Goodale's *Physiological Botany*, attests to its excellence by receiving commendation at home and abroad." The editors noted, for example, that a review in a leading

German journal, *Botanisches Centralblatt,* saw the book "as marking an important event for American science, and [we] rank it in some respects above the text-books of German writers." It also was reviewed favorably in England.[28]

Despite the mixed feelings about the progress of the "new botany," enough change was occurring for some Americans to object to what was happening and to defend taxonomy as where the study of botany should remain. Some of the new botanists' exuberance for their work and their tendency to put down systematic botany as old-fashioned and no longer truly botany drew warnings that a reasonable transition should be made. Defenders of taxonomy feared that in trying to promote the new, the old would be entirely discarded. A writer in the *Gazette,* for example, noted that "as with every reform the first tendency is to swing to the opposite extreme so the new botany begins by abhorring classification." He stated that it "is the case of the superstructure despising the foundation on which it rests, for systematic botany necessarily came first, and if it does not contain all botanical problems, or even the most important ones, certainly it presents some that are extremely important and worthy of any student's consideration." The writer continued that "this is no plea for the study of systematic botany as opposed to the structural and physiological; it is meant to call attention that the pendulum has swung farther away from the old side than it can stay and that the study of botany must include the systematic phase." The writer concluded that "whether the class should begin with Gray's textbooks and *Manual* and then follow with Bessey or begin with protoplasm and run the whole gamut of tissues and tissue systems and then study classification is for the individual teacher to decide, the point is that both should be done."[29]

Although he was often quite critical of the old-style systematic botany, Bessey agreed with this writer in the *Gazette.* His interests in structural physiology increasingly moved to issues and research that related to developing a natural classification system (see Chapter 8). Bessey and the editors of the *Gazette* agreed that systematic botany remained the foundation of their science; therefore, structural and physiological information should contribute in the end to the realization of a natural system. Yet they continually stressed that systematic botany was necessary as a field of study only if the questions asked and the methods of study changed. Systematic botany must become more specialized, they argued, and the basis from which classification was derived must be broadened. They pointed out that even with higher plants, the specialized studies were just making the beginning of a natural system possible.[30]

In 1886 Bessey, in reviewing Coulter's study of the Rocky Mountains, observed that "it is gratifying to note these signs of a recognition, in a systematic manual, of the doctrine of evolution, and of the significance of the structural homologies which are now familiar to every vegetable anatomist."[31] Systematists, Bessey observed at another time, must consider geographic distribution more in their studies. "It is time that botanists bestir themselves in the matter and consider the organization of a regular biological survey that will deal with plants as biological problems and not merely as specimens to be catalogued," he stated, adding that knowledge of "minute" anatomy must be added to the former work in external anatomy to provide a more complete view of plant structure and development.[32]

Thus, while wanting to broaden the study of systematic botany, Bessey was stressing that he wanted a balance—a study of botany that did not restrict itself only to naming plants but likewise one that did not confine itself to the laboratory and the study of cells and tissues. He wanted the "new botany" to retain the older emphasis on the study of actual flora in the field and classification and to add the present trends in structural physiology and the laboratory.[33] As stated in the *Gazette,* "modern methods may claim that they do not want to 'analyze' plants, but they do and always will, and it will remain the chief means of exciting a living interest in botany."[34]

In their concern that botany change, the new botanists realized that the primary basis for change was the method of teaching those students who chose to become botanists. To many of the botanists the test of improved pedagogy was to introduce the laboratory in conjunction with the new textbooks. By the mid-1880s there were laboratories equipped with compound microscopes in sufficient numbers at least for junior and senior classes in a scattering of universities. Harvard had its three well-equipped laboratories, and prominent among the botany departments that had laboratories were those at state universities and land-grant colleges developed by Rothrock at Pennsylvania, A. N. Prentiss at Cornell, Beal at Michigan Agricultural, Volney M. Spalding at Michigan, Burrill at Illinois, Stanley Coulter at Purdue, Barnes at Wisconsin, Douglas H. Campbell at Indiana, and Bessey at both Iowa Agricultural and Nebraska. John M. Coulter had built a good laboratory at tiny Wabash College, and William Trelease was developing an important center at the new Shaw School of Botany in St. Louis. Of course there were other schools that had courses organized on a laboratory basis, but the aforementioned were the ones that were advanced in their ability to provide extensive work in microscopy and served as models for others to follow.[35]

With the growth of laboratories came the need for instructional manuals. Arthur, Barnes, and John M. Coulter, the three editors of the *Botanical Gazette,* led the way with their *Hand Book of Plant Dissection* in 1886, but they were soon followed by others. The *Botanical Gazette,* in fact, labeled this period an "era of textbooks and laboratory guides," and Bessey stated that "one of the significant signs of the times in botany is the multiplication of microscopical and physiological laboratory manuals . . . a few years ago these were unheard of and now we have half a dozen or more."[36] The *Hand Book of Plant Dissection* not only provided exercises for the student but also followed the controversial approach that beginning students were capable of studying and should be introduced to "lower plants."

A strong argument for teaching only the flowering plants to beginning students was that in nearly all cases, even where the college had a laboratory, the beginning students had to make observations and dissections only with the unaided eye or at best with a hand lens. Bessey acknowledged that most of the advancements in teaching botany benefited only the advanced students, and he believed beginners were capable of being introduced to the compound microscope or at least having benefit of it with the aid of the instructor. Even if compound microscopes were unavailable, as was the case in most beginning classes in the United States, Bessey stated that the "old adage, 'a short horse is soon curried,' is appropriate here. One must not expect to see as much in *Protococcus* as in *Capsella,* but one must not neglect to see the little that is to be seen."[37] Thus the average college freshman, who Bessey recognized would probably never take more than the first course in botany, was largely unaffected by the improvements being made in teaching. "The beginner," he stated, "finds himself still obliged to pursue the same old tread mill course with endless lists of bare technicalities and meaningless botanical jargon." Although the subject matter of botany "has undergone a complete transformation," he noted, Bessey nevertheless believed the teaching of beginning botany was nearly the same as when he was a student.[38]

To remedy this situation, Bessey advocated introducing the student to the overall plant kingdom by studying representatives of all groups. The student should collect and study the various groups "not so much for the sake of making specimens as of becoming acquainted with their general appearance, structure, and habits."[39] As Bessey viewed the situation, the problem of teaching only higher plants in beginning courses and avoiding internal structure, which many instructors considered too difficult for the students, was not due so much to a lack of equipment or the complexity of the subject but rather to the shortcomings of the average teacher of botany. Bessey believed most biology teachers were untrained ministers or lawyers "whose ideas and methods are of a generation ago" and whose training does not convey an "idea of modern science." The result was that to this type of teacher "their whole idea of botany is 'analysis' [plant identification]."[40]

As with the attempted emphasis of the "new botany" away from systematics, there were botanists who were wary that teaching methods would likewise swing too far toward laboratory teaching to the neglect of the textbook and lecture. While there was no consensus on how and how much to use the laboratory approach, the topic received increased consideration at meetings and in journals from the eighties onward.[41]

The question of how to improve the teaching of biology was the main force behind the organization in 1888 of the Western Society of Naturalists.[42] Composed of persons teaching in midwestern colleges and universities, the complete program at the first meeting dealt with teaching. Presenting the first paper to the new group, Douglas H. Campbell of Indiana University supported teaching the entire plant kingdom in the "logical" order of starting with the simple organisms and proceeding to the complex. He believed that even high school students could handle the lower orders and that such teaching should make "constant use of the compound microscope."[43] Campbell followed the ideas in his paper in writing a beginning textbook for high schools and colleges that conformed closely to the format of Bessey's *Essentials*.

Although Campbell's ideas coincided completely with Bessey's, they did not fit with the view of many at the meeting; the ensuing discussion revealed a "considerable diversity of opinion as to method, but all agreed in relegating the analysis of flowers to the background."[44] John M. Coulter, for one, was skeptical whether the approach of Bessey and Campbell was possible for high school students and even for college freshmen. Coulter was concerned that in addition to the lack of laboratory equipment in many schools, students were not capable of understanding internal structure before external, or "lower plants" before higher, but primarily he simply believed teachers were not adequately trained in structural physiology to present it properly.[45] Bessey disagreed with Coulter's first objections, but he agreed completely that teachers were improperly prepared and made improved training a prime reform in his own state of Nebraska during the 1890s.

The late 1880s witnessed continued efforts by Bessey and other new botanists to promote professionalism. Journals improved to the degree that the *Botanical Gazette* and the *American Naturalist* varied their coverage to include more nonsystematic botany. Bessey continued to cover a wider range of plant-related topics than did the Harvard botanists who managed the botanical section of the *American Journal of Science,* not so much in subject matter as in promoting professional growth, discussing problems related to teaching, and emphasizing the work of midwestern universities and agricultural botany. Also, the *Naturalist* contained many more articles on botany outside the section of notes written by Bessey. The botany section of *Science* remained virtually nonexistent during the 1880s, however, and disappointingly to Bessey, the new *Journal of Mycology* was basically devoted to listing and describing new species and to family studies.[46]

As in the earlier part of the decade, Bessey and Coulter continued to use their positions with the *Naturalist* and the *Gazette* to convince botanists that the AAAS should be the primary place for setting professional standards. They continued to evaluate the nature and caliber of American research by the papers presented at the meetings and largely judged other professional growth by the activities of the Botanical Club of the AAAS.

Among the highlights, Bessey was enthusiastic about the "excitement" created at the 1885 meeting when his former student J. C. Arthur discussed the bacterial origins of pear blight. The significance of Arthur's methodology and findings would, it was hoped, convince Americans that there was challenging and important work to be done in botany. Yet in calling for papers in 1886 Bessey still pessimistically issued the challenge to "let the effort be continued to rescue botany from the discredit into which it had fallen. Let the charge of superficiality in papers and discussions be made no longer true."[47] Coulter likewise stated that for the "credit of botany . . . the disjointed twaddle which so often offends the section [in the AAAS], if it must be said, should be reserved for the privacy of the Botanical Club, which will guard its family secrets." He called for presenters to write and then read their papers instead of "this thing of talking in a maudlin way for half an hour."[48]

This mixture of pessimism and optimism continued among the promoters of American botany during the late 1880s, but a notable change in attitude regarding the success of the AAAS occurred from 1889 to 1891. Bessey observed that both the 1889 and 1890 meetings had large botany contingencies and were "notable by the activity of the botanical part of the section of biology."[49] Meeting in Toronto in 1889 and Indianapolis in 1890, midwesterners accounted for much of the attention and activity. Bessey believed another important factor was the decision within the Botanical Club to organize a special session in 1890 on plant geography and to have five or six well-prepared and high-level papers. The session received an enthusiastic reception, and those present "ordered" the papers to go into the proceedings in full rather than in abstract and to be distributed in a separate pamphlet. In addition, they decided that the special topical session should become a permanent part of the meetings and designated plant physiology to be the 1891 topic with Bessey, Arthur, Beal, L. H. Pammel of Iowa Agricultural College, and Edmund B. Wilson of Columbia University as presenters.

The botanists believed that other scientists were recognizing the growth and improvements in botany at this time by electing the plant physiologist George Goodale president of the AAAS. Goodale was only the second botanist ever to hold the position, Asa Gray being the first, and to some botanists it was a group more than an individual recognition.[50] The success of the 1889 and 1890 meetings led Bessey to comment that "taken all in all, the botanists of the country have no need of feeling ashamed of the quality of the work done in the Association and the related societies."[51]

While this feeling of confidence and importance that was building was evident both within and without the AAAS, much of the explanation for professional growth was attributed by Bessey and others to the expanding role of the Botanical Club. The club began to provide a setting where botanists could discuss matters of concern only to themselves, and such matters were increasing with specialization.[52] The organized field trips had become the most enjoyable and often most important reason for attending AAAS meetings. But promoters of professionalism wanted the club to have more than a social role. In particular, botanists sought recognition apart from the zoologists and other biologists. Yellow badges, in addition to regular AAAS badges, became the distinguishing mark of botanists at the annual meetings.

Bessey, as chairman of the club in 1885, had recommended a larger professional role for the group. He suggested that the club should provide a forum for discussing and settling issues such as formulating uniform English names of plant diseases and their fungi, adopting uniform pronunciation of Latin names of plants, encouraging the various journals to specialize, and promoting a closer relationship of botanists with the National Herbarium. From that time onward, several committees each year worked on these and related issues. Occasionally the club forwarded resolutions to the general business meeting of the AAAS where it was hoped the entire organization could put its support behind such items as creating a full-scale botanical division within the USDA and establishing a National Arboretum.[53]

In recognizing the improvement in botanical papers presented to the AAAS and the increased attendance, the editors of the *Botanical Gazette* credited the change to the influence of the club. Discussing the growing strength of American botany, the *Gazette* claimed that "it is probably safe to say that the botanists form the best compacted organization of scientific workers in the country." Continuing, the *Gazette* explained, "Never before has there been such botanical activity in this country, and no small part of the cause is due to the botanical journals which supply the means of speedy publication, and the meetings of the Botanical Club, which brings all workers into a more sympathetic relationship."[54]

Growth of the club had been rapid since its creation in 1883. In 1885 eighty persons attended its meetings, double the number of botanists that had been at any previous AAAS, and in 1887 attendance was 140. Not all who attended the club meetings were botanists, however. Of eighty-five who registered in 1885, only thirty-seven listed botany as their primary field. Still, by 1890 botanists constituted nearly one-fourth the total attendance of the AAAS and 65 percent of those in the Biology Section. Even more revealing, attendance at sessions of the Botanical Club was larger than those of the Biology Section. Earlier the club had squeezed into time periods when the Biology Section was not meeting, but as its business and papers increased, complaints arose that there was not enough time for the club. In 1885 the club held six one-hour sessions, but by 1891 sessions extended to three full days and the following year ran "almost continuously." No longer was the club merely an adjunct of the Biology Section.[55]

In light of these developments, as early as 1886 members of the Botanical Club discussed creating a separate section of botany within the AAAS. Shortly the club established a committee to pursue separation, and in a related matter the botanists asked the council of the AAAS to publish the proceedings of the Botanical Club as part of the annual report of the association. Although more and more botanists were skipping the zoology papers in the Biology Section to spend time with the Botanical Club, opposition to a separate botanical section remained strong during the 1880s. There were many teachers and amateurs who had not developed a specialty and who felt out of place, socially and intellectually, among the new botanists. The editors of the *Botanical Gazette* vigorously opposed separation and argued that a separate botanical section would destroy the club, which they believed played such an important and unique role in professional growth. Also, there were very significant groups such as the physiologists and bacteriologists who had not yet completely distinguished their identity as either botanical or zoological.[56]

With successful AAAS meetings in 1889 and 1890 most botanists accelerated the move for a separate section. The botanists now dominated the Biology Section in both numbers and papers presented. The role of the zoologists was rapidly declining, and many were shifting their alliances to organizations other than the AAAS.

The Biology Section at the 1891 meeting in Washington, D.C., was even more crowded with botanists. The Botanical and Entomological clubs met "almost continuously," the preplanned session on plant physiology—including papers by Bessey and his student Albert F. Woods—proved as successful as the previous year, and in general, botanists were exceptionally active in related societies meeting in conjunction with the AAAS, such as the Society for the Promotion of Agricultural Science and the American Microscopical Society. This time when the matter of separation from the zoologists and biologists was raised in the club there was no opposition.[57] Even the influential editors of the *Botanical Gazette* capitulated when confronted with the sense of achievement and independence generated during the past several meetings. Not only was there insufficient time for all the botanical papers, but also as the *Gazette* admitted, specialization had occurred to the point where many papers of the one group were no longer intelligible to the other, and it was "apparent that one half does not care how the other half lives." There were to remain some sessions in which papers of dual interest were apparent, and the botanists voted to continue the Botanical Club even after separation.[58]

While Bessey and Coulter concentrated most of their efforts on the AAAS, there were other organizations that attracted botanists. Midwesterners had few organizations aside from the newly formed Western Society of Naturalists and the emerging state academies of science; therefore, they more easily concentrated their energies in the AAAS. Many eastern botanists, however, although not alienated from the AAAS nevertheless had loyalties to the numerous local and regional organizations, most of which were long-established and prestigious.

Aside from the AAAS, much of Bessey's interest in promoting institutional growth was directed to the USDA. Increasingly the growing prominence of agricultural botany, particularly in land-grant colleges and experiment stations, shifted much of the institutional growth in that direction. Although botany had been included in the national surveys, that role was declining and the USDA was a natural institution to look to for growth, funding, jobs, and recognition. Bessey, through the *American Naturalist,* continued to promote the USDA, and the *Botanical Gazette* increased its coverage of agricultural topics. Yet while zoologists and particularly microbiologists were having success in the growing scientific role of the USDA, botany was not prominently featured. Though remaining optimistic, Bessey and the *Gazette* were disappointed overall with botany in the USDA during the 1880s both in terms of relative money spent between botanical and zoological problems and in terms of the kind of results that came from the department. Not only were such problems as plant diseases receiving much less attention than animal ailments, but what research was being done Bessey and the *Gazette* editors judged to be of poor and generally useless quality to both scientist and agriculturist.[59]

In other agriculturally related organizations, Bessey and other botanists likewise pushed to expand their professional and scientific interests. The Society for the Promotion of Agricultural Science (SPAS) in particular had a "strong botanical coloring." About one-fourth of the members were botanists. Although this faction was second in number to the chemists, in terms of activity the botanists surpassed all other disciplines. Botanical papers presented at the annual meetings were by far the most numerous, and active leadership by botanists was evident. In 1889, for example, the year that Bessey was SPAS president, all the officers were botanists. Although the SPAS remained a hope of professionals interested in advancing agricultural science, membership and attendance remained small, and for most members its activities remained secondary to their interests in the AAAS, to which the SPAS tied its meetings.[60]

Another closely related organization was the American Association of Agricultural Colleges and Experiment Stations (AAACES). Before 1890 this organization had little appeal to scientists, but in 1889 led primarily by William J. Beal of Michigan Agricultural College, thirteen experiment station botanists met while attending the AAAS and discussed some of the problems associated with their positions. They pondered such issues as the nature of station bulletins, whether bulletins should be for popular reading or be reports on experiments and research of interest primarily to specialists, whether bulletins should contain only original work or pass on any information considered valuable to farmers, how exchanges of specimens could operate, and the matter of having yearly meetings. On this last item, a committee of three, including Bessey, was to discuss the question with the director of the Office of Experiment Stations and to arrange for separate scientific and administrative sections for the next meeting of the organization. Within the AAACES there was opposition to the organization subdividing by scientific disciplines, but finally at the 1890 meeting an agreement was made to create five sections.[61]

Although there were twenty-nine botanists attached to experiment stations in 1890, attendance at the AAACES, like the SPAS, remained small. Bessey, for example, attended only when the AAACES met in conjunction with the AAAS. In 1889 he did send a paper to be read on the advantages of having the experiment station closely tied into the teaching and research of the university and its regular science departments—the plan he was attempting to establish at the University of Nebraska.[62] The 1891 meeting of the AAACES in Washington, D.C., demonstrated how thinly station botanists had spread themselves. Dividing their time among the Botanical Club and Biology Section of the AAAS, the SPAS, and the botanical section of the AAACES, the latter two fared poorly. For most station botanists at this time, agricultural interests, although considered important, were secondary to their main professional and scientific interests.[63]

While the 1880s were a period of growth and a time of optimism for professionalization of American botany, a feeling of uncertainty always remained. While seeing progress on the one hand, botanists saw on the other hand that much of the gain was loosely established and often more illusory than real. They still viewed the fulfillment of the goals of professionalism and the "new botany" as yet ahead of them. Still the journals needed to improve, the institutional structure had weaknesses, good teaching and research were restricted to a small segment of the total number of botanists, and there was almost nothing gained in the area of financial support. While recognizing these shortcomings, most promoters of professionalization and the "new botany" nevertheless believed they were on the verge of great achievements in the United States.

# 5
# Professionalization and Reform

THE INTENSIFICATION of interest in professionalization that was felt so strongly in the American Association for the Advancement of Science (AAAS) meetings of 1889–1891 did not subside during the remainder of the 1890s. It was in light of this background that botanists went to Rochester, New York, in 1892 anticipating another fruitful meeting which would further spotlight the development of American botany. The number of botanists at Rochester was proportionately large. In particular, botanists came from the Midwest and from the governmental agencies in Washington, D.C.; there was a noticeable lack of attendance from New England and, as usual, almost no one came from the South and the Far West.[1]

Botanists, zoologists, and other biologists met for the last time together in the Biology Section, and although Bessey was unable to attend the Rochester meeting and to help in the reorganization, he had made his ideas known prior to the meeting. Bessey had opposed earlier attempts to divide the Biology Section, but he believed the botanists were "now strong enough to go alone." He desired to maintain the Botanical Club and return it to its original role of being a place for shorter and less important papers and as the place for social functions and field trips. Most of the professional concerns that had become the major business of the club in recent meetings they could now conduct in the new Botany Section.[2]

A number of issues filled the time of the club, so much so that little time remained for reading papers. Several pressing matters that had greatly interested Bessey were establishing uniformity in pronunciation of botanical terms, requiring metric measurements in all botanical descriptions and writings, and obtaining tariff reductions on foreign scientific equipment and publications.

These issues remained before the botanists during the next few years. Bessey delivered a short talk to the club in 1893 on pronunciation reform in which he supported Roman versions and was appointed to chair a committee "to stir the matter up."[3] Regarding metrics, Bessey often used book reviews to plea for adoption of the decimal system, but his concern for this reform reached a peak in 1897 with the publication of two volumes on major American floras, Nathaniel Britton's and Benjamin Robinson's, which both used "these miserable feet and inches." In his letters to both, Bessey noted that "I really worked myself up into a fury over them and have a bash against that kind of thing in the February *Naturalist*. I hope you will knock out every one of those old measurements and substitute metric units." Not only did Bessey believe metrics were easier to use and more uniform with foreign botanical publications but he also thought that professionally botanists should be embarrassed by being last among the sciences to change.[4] As did others, Bessey pushed his representatives in Congress for duty-free scientific material, particularly microscopes and journals. Confident of success, Bessey was appalled by the Dingley tariff of 1897. "We ask for bread and he has given us a stone," Bessey exclaimed; "I have been a member of the Republican party for all my life and I don't like to see its servants in Congress lead it towards destruction by such trickery, for trickery it is, and nothing else."[5]

While a number of these professional and botanical issues were discussed at the 1892 meeting, the topic that had most attracted the large entourage of botanists to Rochester was the previously announced discussion on nomenclature reform. This matter had been brewing for quite some time. Among systematic botanists in America, although there were always differences of opinion associated with the naming of plants, the situation became particularly unsettled in the 1880s when the California systematist Edward Greene challenged the philosophy and methodology of Asa Gray. In particular, there was a basic difference between them regarding the handling of earlier-named species with Greene preferring to combine a number of old species into one. There also was a disagreement about which name should be used when more than one had been given for the same plant. Which was the correct name and which was the synonym, and whose name should be attached as author? Gray tended to discount old and what he considered incorrect or incomplete names, and he renamed them believing the first correct combination of generic and specific names had precedence over half-names (where the species was in an incorrect genus). In this and in most practices of nomenclature, Gray followed the lead of the botanists at Kew Gardens in England, particularly Joseph Hooker. Greene adhered to the strict law of priority, and he followed the practice of establishing an initial starting date and of accepting the first specific name applied to the plant after that date.

In practice Greene would not accept many of the name changes of the nineteenth century by such botanists as Gray and Hooker if he could find an earlier specific name for the plant in question. As a result, persons sharing Greene's approach—following the rule of priority—favored discarding a number of plant names because they found an earlier published name. Sometimes these names had a century of usage. To Gray the practice of strict priority was letting rules dictate to good sense, and to Greene it was placing nomenclature under the "rule of laws" rather than under the arbitrary control of a few persons such as Gray and Hooker.[6]

Until his death in 1888 Gray was generally successful in discrediting Greene and holding the line against those termed the "radical" reformers. Sereno Watson carried on for Gray until 1892 when the "Cambridge principles" became the responsibility of the systematic botanist Benjamin Robinson.[7] In the United States, therefore, questions that arose under the International Code of Nomenclature, last formulated at Paris in 1867, had been interpreted and applied largely by Gray and a few other general botanists.[8]

Much of the desire to bring the issue of nomenclature before the Botanical Club had been stimulated by Nathaniel Britton, a young botanist from Columbia University and later the New York Botanical Garden. Britton basically agreed with Greene regarding principles of nomenclature, and in a paper to the Biology Section of the AAAS in 1888 he emphasized the faults in the current international code. With Gray now gone from the scene, and with the increase of specialists, interpretations and opinions about proper nomenclature multiplied to the point that the number of botanists who agreed with Britton on the need for reform was sufficient to make it a special topic for the 1892 Rochester meeting.[9]

Bessey agreed with Britton that changes were necessary to secure uniform practices in nomenclature, but before the Rochester meeting he warned that "such action must be cautiously taken." He suggested some papers on the subject to clarify the issues and the appointment of a committee. "It will do no good to try to settle any points by a vote taken now," Bessey surmised, "the matter must be thoroughly discussed before hand and the time of a vote fully advertised." Bessey was thus largely prepared for the action taken at the Rochester meeting.[10]

Once the meeting was in session Britton initiated the consideration of nomenclature reform with a motion to appoint a committee to formulate a set of proposed rules for the consideration of the club, and this

## PROFESSIONALIZATION AND REFORM

was done immediately. As chair of the club, Britton appointed himself, John M. Coulter, H. H. Rusby, William A. Kellerman, Frederick Coville, Lucien Underwood, and Lester Frank Ward to the committee. This group quickly finished its work, and about twenty-five botanists worked over the committee proposals before they were placed before the Botanical Club. Those present voiced little opposition to the proposed reforms, and of the proposals, generally referred to as the Rochester Rules, only one had enough opposition to generate a motion to change it and the motion lost. With the approval of the reforms by the club, Britton next moved to establish a permanent committee that would provide a practical demonstration of the new rules by using them to compile a Checklist of Flowering Plants in the geographical area covered by Gray's *Manual* (6th edition), with the addition of Kansas, Nebraska, and Canada. This motion carried, and the committee that had formulated the Rochester Rules was made the permanent committee in charge of the checklist. The committee members were to receive all comments on their work and report these opinions to the next meeting of the club.[11]

In a closely related item of business, club members discussed the appointment of a representative to the International Botanical Congress in Genoa, Italy, the following month. Again it was Britton who successfully pushed for Lucien Underwood, a strong proponent of nomenclature reform and a member of the checklist committee, to represent the AAAS. Underwood of De Pauw University was to present the Rochester Rules to the Genoa congress as the American position on reform.[12]

In other significant discussions during the Rochester meeting, the botanists considered the feasibility of establishing a new society and of hosting an International Botanical Congress in conjunction with the Chicago Columbian Exposition in the upcoming year. While the idea of a new, elite society that would recognize the advancements taking place in American botany met with general favor, the consensus was that such an organization would be unsuccessful at present. Also, those in attendance at Rochester were in general agreement that most Europeans were not yet ready to attend a congress in the United States. Nevertheless, the botanists, with Bessey chosen as a vice president and thereby chair of the new Botany Section of the AAAS, looked forward to their 1893 meeting again being special.[13]

Thus botanists at Rochester displayed a mixture of confidence and caution, and in the proposals for revised rules of nomenclature, a new society, and an international congress, they were divided on the immediate course of their profession. There was little disagreement about building upon the general plan of the "new botany" as formulated in the 1880s, but some leaders believed that the progress they had envisioned during the previous decade had in large part been achieved or was well underway. Having achieved a real or at least a symbolic plateau, it was now time to move boldly onward. This group, of which Bessey was one of the leaders, was optimistic and was enveloped in the sense of achievement that was, or at least seemed to be, taking place in American botany. Hence, they were anxious to display this achievement by having foreign botanists come to the United States and witness the developments for themselves, by showing that American systematists had the best solutions to the complex nomenclature questions and should share equally with Europeans in arriving at a settlement of a reformed code, and by further recognizing those persons who were responsible for the growth taking place in the form of a new society. National pride was manifested by some, such as Underwood, who upon his return from Europe trumpeted: "We have deferred altogether too much to Kew in regard to our plants!!!! . . . I am going to touch up the British spirit a little when I get a chance. I feel more proud than ever of our American botanists after seeing the English."[14]

Other leaders were more cautious and less certain of the actual achievements that had taken place. They wanted more consolidation of the current gains before moving into new or more advanced professional ventures. They were afraid that the assertion of American leadership in the nomenclature issue, hosting an international congress, and establishing a new society were premature, and failure in these ventures at this time would jeopardize the gains that American botanists were making. Changes also would disrupt the traditional structure of American botany, and there was uncertainty as to what the results would be. Despite these disagreements over the time schedule and particulars for the progression of professionalism, as an editorial in the *Botanical Gazette* proclaimed: "The Rochester meeting bids fair to mark an epoch for American botanists."[15]

Having completed what they judged their most successful meeting and while preparing for what they anticipated to be an even more eventful gathering in 1893 at Madison, Wisconsin, the issue that continued to generate the most attention was that of nomenclature and the Rochester proposals for reform. At the International Congress in Genoa, Americans presented the Rochester proposals for consideration, but the congress favored recommendations formulated by a group from Berlin that covered only the issue of genera. It did, however, substitute the American date of 1753 (publication of Linnaeus' *Species Plantarum*, 5th ed.) as the starting point for recognizing legitimate species names. Underwood, representing the AAAS, stated that the Berlin proposals were not significantly different from their own, and he expressed his satisfaction that the nomenclature questions were "practically settled." In line with his optimism, Underwood stated that "I believe more than ever in the power of American botany to swing the whole business as we think best if we go a little slowly and remember that ideas come to most Europeans by freight and have to be absorbed by capillary action in homeopathic doses."[16] Recognizing that in light of specialization one set of rules could no longer anticipate all the questions of nomenclature, the Genoa congress appointed an International Committee on Nomenclature whose members—including the Americans Nathaniel L. Britton, John M. Coulter, and Edward Greene—were to handle all questions and problems on a continuing basis and to suggest revisions in the International Code.[17]

In the meantime, the idea of hosting an International Botanical Congress in the United States in 1893 had not died despite the strong opposition voiced against it at Rochester. J. C. Arthur in particular continued to lobby for the congress, and he was reinforced by Underwood who came back from Genoa stating that the Americans must hold a congress because the "time was ripe for it" and that European botanists expected it and would attend. Particularly with the progress made at Genoa, Underwood believed a congress would allow the International Committee on Nomenclature to meet and continue its work toward a stable and standard set of rules. At this point, Bessey, as chair of the upcoming meeting of the AAAS, announced that it was evident that they should hold an international congress in the United States. Not having any permanent organizational machinery to handle such a situation, Bessey took it upon himself as sectional chair, along with the chair of the Botanical Club, to appoint a committee to plan the congress.[18]

The editors of the *Gazette,* feeling Europeans would not attend, had opposed the congress, but immediately before the Madison meeting they also now enthusiastically sounded the call for the congress. They explained that the nomenclature issue should not be delayed and that nationalism and professionalization in America required the meeting. "America needs to be botanically discovered by Europeans," they proclaimed. "The country has botanical wealth; a little crude, it may be, when compared with foreign riches, but still there is wealth, and it has yet barely come to the notice of most foreigners." Still somewhat apprehensive regarding the success of the congress, the editors concluded that "if the present congress does not prove all that its well-wishers could desire, it may yet be a means of eventually securing on American soil the truly representative international congress to which all will be willing to concede authority."[19]

As time drew closer to the Madison meeting, articles and letters appearing in the journals demonstrated that the nomenclature issue had been anything but settled at Genoa. In the first place, a number of botanists in both Europe and the United States challenged the authority of the Genoa meeting to bind all botanists because sixty of the one hundred attending were Italians. Secondly, the Berlin proposals did not cover all disputed points. Finally, questions continued about what was to be done with the Rochester Rules and the Checklist of Flowering Plants that the Americans were preparing.[20]

One of the leading critics of the Berlin rules was the German botanist Otto Kuntze. His writings had helped generate feelings favoring reform prior to the Genoa and Rochester meetings. He had bombarded the United States with letters and writings, and Bessey had served as one of his intermediaries in distributing his ideas to Americans. In addition to opposing the work at Genoa, Kuntze criticized Americans for pursuing their Rochester Rules, and the German raised a number of questions that became centers of discussion at the upcoming meetings in 1893. Were the ideas formulated on nomenclature at Rochester proposed reforms or rules? Did Americans intend to use the Rochester Rules before or without international agreement? Kuntze raised these questions particularly in reference to the Checklist of Flowering Plants.[21] The charge by Kuntze that American reformers were acting nationalistically was supported by a group of Americans led by Benjamin Robinson and Erwin F. Smith, along with the Canadian W. F. Ganong, all of whom claimed that the reformers did not even represent the dominant American view.

As a "radical" reformer on the nomenclature issues, Bessey was extremely impatient with those botanists he derisively termed "conservatives." These were persons who he believed always opposed changes—no matter how sound in principles—on grounds that to change terminology, rules of nomenclature, or any long-established usages would cause confusion. Bessey contended that to continue outdated usages in plant names was more harmful to botany than the confusion which undoubtedly would come from new procedures in identifying and naming plants and from the changes in old to new names. He felt the best way to bring about changes was to put the reforms in use and to demonstrate in this manner their benefits. Past authorities, Bessey believed, should never control the present and keep improvements from being introduced into science. Although the principles and reasons for procedures by past authorities were good for their day, he conceded, nevertheless the rapid transformation of botany required that rules and codes change to serve the usefulness intended. Specialization and the increasing number of botanists rendered the work of past authorities more and more quickly out-of-date. Bessey therefore, although respectful of the accomplishments of his botanical predecessors, believed that authorities such as Carolus Linnaeus, whose publications in the 1700s marked the beginning dates of most nomenclature codes; Alphonse de Candolle, the primary author of the Paris Code of 1867; and Asa Gray, the past interpreter of American nomenclature; were not as qualified as the current professionals on most questions of modern botany. Having missed the Rochester meeting, Bessey made his influence evident before and during the Madison meeting, where he was a leader in the debates and where he presented his ideas concerning the checklist, decapitalization of all species names, and not using personal names when naming species.[22]

As in the previous year, botanists went to the 1893 AAAS in large numbers and with enthusiasm to prove to other scientists that they deserved to have their own section separate from the zoologists. Bessey opened the new Section G with his vice presidential address of about one hour on "Evolution and Classification" in which he pointed out the limited degree to which concepts of evolution had been applied to plant classification (see Chapter 8). In their new section, botanists presented about one-fifth of all the papers at the Madison meeting.[23] Bessey's program committee had attempted to arrange for most of the leading new botanists to be on the program, and they had tried by preplanning special sessions to have a wide representation of all the various areas of botany, thus displaying the caliber and diversity which American botany had attained. Bessey, however, had some difficulty covering all areas of botany, noting that "we are still mostly systematists after all our talk to the contrary."[24]

One of the sessions planned by Bessey's committee that generated considerable response featured his former student Conway MacMillan of the University of Minnesota. Presenting some of his observations on laboratory and classroom teaching in American colleges, MacMillan stressed that a wide divergence of techniques still prevailed and that too many teachers still did little more than the "analysis of a few flowers" using Gray's *Lessons*. The discussion that followed the paper resulted in the adoption of a resolution to the U.S. Commissioner of Education requesting a published report on botany teaching to be prepared by MacMillan. In addition, the section named a committee to report the next year on how botanists of the AAAS might use their "influence" to encourage more "new botany" in the high schools. In one of the other special sessions, Nathaniel Britton presented a summary of the nomenclature controversy.[25]

Bessey was extremely pleased with the first meeting of the new Botany Section, concluding that this was the "most notable gathering of American botanists in recent times."[26] Despite the satisfaction with the new section, as usual most of the activity for the botanists continued to take place in the Botanical Club. As expected, the first item of importance was the report of the committee that was to prepare the Checklist of Flowering Plants in accordance with the Rochester proposals. The committee reported the checklist ready and recommended its publication following the amendment of several minor points in the Rochester proposals that came to light through their experiences in drawing up the document. There was enough disagreement and uncertainty about the checklist that members of the club voted to send it back to the committee for one final reconsideration. The committee returned the next day, its members again asserting that they wanted to make no further changes in their original report. The proponents of reform were dominant enough in the club to prevent any further alterations, and a vote authorized the committee to proceed with the printing of the Rochester Rules and the checklist.[27]

Unlike the previous year, some of the opponents of reform—or at least of reform as embodied in the Rochester Rules—expressed their opposition. Much of their disapproval centered around the undetermined status of the Rochester proposals. As had Kuntze, they asked if the ideas formulated at Rochester were proposals or rules. Did the reformers intend to use the Rochester Rules without international agreement? The opponents argued that to change terminology, rules of nomenclature, or any long-standing usages would cause more confusion than benefits.[28]

Radical reformers defended proceeding with the checklist and with other publications using the Rochester Rules by stressing basically the points that Bessey had made regarding not being bound by past authorities, the need to change more rapidly with increasing numbers of botanists and better-trained ones, and that to continue outdated usages in nomenclature was harmful to the development of botany. They believed the best way to bring about changes was to use the rules and demonstrate the benefits.

The other major item of business the club carried over from Rochester was the report by the committee for the proposed new botanical society. The majority of the committee, representing eight of the ten members, had not changed their opinion that the creation of such a society was premature. Charles Barnes, however, presented a minority report that declared that there was no more opportune time than the present to found the new society. He argued that the Botanical Club was open to all interested botanists and that another society, restricted in membership to only active researchers, would greatly enhance American professionalization. He proposed that those present elect ten persons who would in turn select fifteen additional botanists, and these twenty-five would constitute the charter membership of the American Botanical Society. Ignoring the majority report, two-thirds of the club members accepted Barnes' proposal and following a long discussion proceeded to elect the first ten charter members.[29] Bessey was among them, and he met with the others to make temporary plans for the society. They determined to meet the next year in conjunction with the AAAS and directed five of their number, headed by William Trelease of the Missouri Botanical Garden, to prepare a constitution for their approval.[30]

With the close of another successful AAAS meeting, many of the botanists stayed on in Madison for the International Botanical Congress. Much of the enthusiasm quickly subsided, however, when few foreign botanists appeared among the thirty-seven persons present, and the meeting was demoted from an international congress to merely the Madison Botanical Congress. Not being international in scope, those present voted that they had no authority to act on the nomenclature issues and the major reason for their meeting was not discussed. Nevertheless, those present proceeded to take advantage of the meeting to try to settle some of the many problems confronting botanists. Five committees of three persons each set to work on the need for and means of establishing a bibliography of North American botany and on the necessity of clarifying terminology used with plant diseases, anatomy and morphology, physiology, horticulture, and geographical botany. After reports from these committees and discussion, the congress voted to enlarge some of the committees, making them standing committees within the AAAS and instructing them to continue their work until the next annual meeting.[31]

Bessey was not on any of the major committees. His involvement in the Madison congress was limited to membership on the nominating committee, helping to write a resolution commemorating the death of George Vasey, a longtime botanist for the USDA and curator of the National Herbarium, and drafting a letter to the U. S. Congress warning of the danger to the National Herbarium if it did not provide a fireproof building for the collections. Finally, when enlarged and made a standing committee, the committee on the nomenclature of plant diseases added Bessey to its rolls.[32]

Certainly there was disappointment with the Madison congress. Foreigners had largely ignored the first attempt by Americans to host an international botanical meeting, and with the failure of the International Commission on Nomenclature to convene, the systematists were unable to clarify the status of the Rochester Rules and the Checklist of Flowering Plants.

To promoters of American botany, however, pride was more evident than disappointment, and they outwardly emphasized only the successful aspect of the nearly two weeks of meetings. They judged the Madison gatherings as more productive than the one in Rochester. Bessey, for example, stated that the concentration of active researchers and professional botanists was noteworthy, as was the prominence of young botanists whom Bessey saw as the first products of the "new botany" movement.[33] To the editors of the *Botanical Gazette,* the success of the Madison meetings was most evident in the feelings of nationalism, unity, and professional pride that were demonstrated by the fact that so many of America's leading botanists met and worked together for two weeks on some of the most troublesome problems facing their scientific discipline. "We should have had no occasion to be ashamed had Germany and England and France been well represented," concluded the *Gazette.*[34] It was the growing cooperative nature of American botany—as represented in the work of the Botanical Club, in the Rochester Rules and the checklist, and in the Madison congress—that so impressed the editors of the *Gazette.* They believed that the day of the individual in botany was ending, and specialization required cooperative general studies, such as floras, and made common terminology and nomenclature even more essential than in the past. Other scientific disciplines had faced this stage in their development, and now botany was confronted with it.[35]

The Rochester and Madison meetings provided abundant groundwork and controversy for the ongoing development of American botany during the remainder of the 1890s. With the exception of 1895, meetings of the Botany Section and the Botanical Club of the AAAS continued to draw good attendance, and the promoters of professionalism approved the quality of the papers. Of the standing committees on terminology established at the Madison congress only two, bibliography and plant disease, made reports at later meetings of the AAAS, and only the former proved of lasting value. The bibliography committee in its first year completed an author catalog of American botanical writings and published it in the Torrey *Bulletin.* With this start the committee continued on a regular basis until its work was taken over by the Torrey Club in 1900.[36]

The most immediate legacy of the Rochester and Madison meetings, and the one that posed the greatest threat to unity and to the objectives of the promoters of professionalization, remained the controversy regarding botanical nomenclature. While the publishing of the Checklist of Flowering Plants was supposedly intended to provide a practical test of the usability of the Rochester Rules, no definite determination had been made to sanction the rules for general use in the United States.[37] Even the reformers who supported the rules and the checklist were divided as to whether the checklist was "simply an expression of principles already adopted" or whether the list was "prepared for the purpose of furnishing a basis of discussion of the principles" of the Rochester Rules.[38] Following the 1895 meeting of the Botanical Club, editors of the *Botanical Gazette* supported the second position—that the principles were not yet accepted. Their thinking was that the nomenclature committee was retained for the purpose of continuing to formulate proposed principles which would then be presented to an international congress. Despite this interpretation, the *Gazette* editors acknowledged that there was no clear agreement and that some American botanists regarded the issues of the Rochester and Madison meetings as a "settled and formal expression of the Club." Certainly the "Rules for Citation" issued by the Madison congress had been published with the appeal that "writers and publishers of botanical matter are earnestly requested to adopt forms here recommended." As the *Gazette* pointed out, "the real difference of opinion seems to be the advisability of using such a code as a guide in publication before it has been sanctioned by an international congress and this question ought to be kept distinct from the approval of the reform principles themselves."[39]

Individual statements of opposition to reform, and to the particular reforms in the Rochester Rules, appeared in the journals, but the most organized and comprehensive attempt to block the rules and the checklist came in the form of a statement accompanied by seventy-four signatures, referred to as the Harvard Circular of Protest. The basic contention of the circular, written mainly by Benjamin L. Robinson of Harvard University, was that the action of the Botanical Club did not constitute any official sanction because many of the persons who voted on the Rochester Rules were not authorities on nomenclature. Besides, he claimed, American botanists had never officially granted authority to the club and the AAAS to speak for all botanists. Certainly outside the United States there had been no official approval of the Rochester Rules. Even unofficially, Robinson argued, the leading international authorities on nomenclature opposed the suggested reforms and viewed them in conflict with the Paris Code even though some American reformers claimed they were not.

As far as their opposition to the principles of the Rochester Rules, the signers of the Harvard Circular thought the changes were unnecessary and would result in tremendous confusion. They believed that once applied the rules would require changing many old and familiar plant names and that the uniformity achieved would not be worth the resulting problems. Also, Robinson emphasized, unlike zoology in which similar reforms had already been adopted, botany was used by many related disciplines such as horticulture, agriculture, and medicine, and all changes in plant names would affect these areas. The changing of familiar names would be nothing but a nuisance to persons in these related areas particularly since they would not be familiar with the basis for the changes made by systematic specialists.[40] Because of his strong objections to the Rochester Rules and to the persons responsible for them, Robinson repeatedly refused his election to the nomenclature committee that prepared the checklist.[41]

Numerous reformers responded immediately to the Harvard Circular. Foremost was Frederick Coville of the USDA who argued that the use of the Botanical Club to formulate the Rochester Rules had been the best way to assure widespread participation, and he believed the reforms would achieve the desired stability in nomenclature. Convinced of propriety and soundness in their cause, Coville therefore believed the protesters were motivated largely by "personalities rather than principles." As Coville expressed it privately to Bessey, "a hard fight is being made . . . a fight based largely on personal grounds and the argument widely used is that the whole code is the outcome of a personal attack by Dr. Britton on the Harvard Herbarium and its work." He concluded that "the idea as you undoubtedly recognize is entirely incorrect, the movement being a general and inevitable movement so far as most of us at least are concerned."[42] Having answered Robinson in the pages of the *Botanical Gazette,* Coville pleaded with Bessey to follow in the *Naturalist:* "A statement of your views will be important in allaying the fears of some of our botanists who appear frightened by the appearance of the Checklist."[43]

As previously noted, Bessey had immediately accepted the Rochester Rules and had recognized them as the basis for necessary reform. Regarding reform, Bessey saw Edward Greene as more "radical" than himself but he believed Greene's views were "more nearly right than those of his critics." Greene was "inclined to be ultra-radical in these matters," Bessey observed, "yet his state of mind is more hopeful than that of the people who propose to keep everything just as it is now." Bessey continued that "we ought to be radical enough to *keep moving* . . . don't wait until you are compelled to move by others . . . be a locomotive not a freight car, that is what Prof. Greene is."[44]

Bessey had been enthusiastic about the compiling of the Checklist of Flowering Plants and provided Britton with a number of his ideas and corrections as the list was in the process of being made, and along similar lines he made numerous suggestions for Britton's *Systematic Botany of North America* when it was in its formative stage. When discussing with Britton the need for some good, "strictly modern work" on fungi, Bessey lamented that J. B. Ellis and Charles H. Peck, two of America's leading authorities, needed to be "made-over." "Why can not these men who know so much look at the vegetable kingdom through modern spectacles?" Bessey asked.[45]

Bessey wrote a review in the *American Naturalist* after the publication of the checklist and acknowledged that while some would oppose it the "book is the mark of a revolution . . . the book is a sign that the day of authority as such is ended and the day of law has begun; the day of botanical equality before the law has come and the humblest botanist now may lawfully correct the greatest."[46] Bessey also reviewed Britton and Addison Brown's *An Illustrated Flora* which used the Rochester Rules, and aside from their not using metric measurements, Bessey was enthusiastic about the work. In thanking Bessey for his "charming review," Britton reported that Theodore Holm of the U.S. National Museum, one of the signers of the Harvard Circular, also was writing a review of his book that "looks a bit as though there was going to be another view of the thing! Blackguardism I suppose."[47]

Bessey responded to Frederick Coville's request for a public statement against the Harvard Circular with a letter to the nomenclature committee of the Botanical Club and then published his views in the *Naturalist*. While agreeing with several provisions of the Harvard protest, such as not changing long-established family names, he explained that despite individual disagreements, which everyone must have with any set of rules, the reforms would achieve the desired stability. In particular he defended the law of priority for generic names. He believed that making exceptions of any kind to this rule would defeat stability.[48]

Continuing, Bessey characterized the statements in the Harvard Circular as unclear and confused. Also, he believed that "much of what is said does not apply to the Rochester Rules." Bessey then discussed each point of the circular, specifying what he felt were misunderstandings of the rules by the protesters and what was faulty thinking, as in the case of not wanting the synonym rule to be retroactive. In concluding, he stated that the protesters exaggerated the number of names that would have to be changed under the reforms, and he thought they were hiding their real motive—which was a basic difference in philosophy that preferred individual judgment in nomenclature over rules—behind "nit-picking" arguments that in too many cases were not "progressive."[49]

Although Bessey's public statements followed his mild but to-the-point approach, privately he was caustic toward the nomenclature conservatives, or as he called them, the "reactionists." Early in the controversy Bessey had expressed the need for reform and his opposition to the dominance of Asa Gray over nomenclature. He believed Gray's *Manual* was no longer a good authority on nomenclature for writers to follow, and he noted that revisions of the *Manual* "ignored many changes in nomenclature which recent investigations had shown to be imperative."[50] Apologizing for not being able to attend the AAAS in 1895, Bessey notified Britton that he was "sorry, as I wanted to be in the fight," and he jabbed at the conservatives with the observation that in supporting and using the Rochester Rules at the University of Nebraska "we are not attacked by 'brain-bog' because of the extra strains which come to us on account of the adoption of the names in the Checklist." He concluded these remarks with the observation that "some of the brethrens [conservatives] appear to be suffering."[51]

As usual, many of Bessey's opinions surfaced through his students. Albert Woods attacked Erwin F. Smith, USDA pathologist, for using the section on plant physiology of the *American Naturalist,* of which he was editor, to support the Harvard Circular.[52] Another student, R. Kent Beattie, apparently was repeating words learned from his mentor when upon meeting Aven Nelson of the University of Wyoming he wrote Bessey that Nelson was "up to date in spite of the fact that he went to Harvard a year. He believes in the Rochester Rules."[53]

Although Bessey could be harsh in his judgment of the nomenclature conservatives, he likewise came in for some heat himself. Marcus Jones, for example, a collector in the American West and a bitter opponent of Greene, wrote Bessey that "I have given up all hope that you Brittonian partisans will be fair, you are driven to the last ditch, you are dead already and like the Irishman's turtle you are dead but not sensible of it." He continued that "I suppose you will continue to wiggle your tails till the sun goes down. Never mind I shall have you on the gridiron as soon as I get time to finish my article on nomenclature and shall be able to roast you and still be fair, so fire away."[54] The nomenclature controversy indeed remained hot.

Going beyond basic differences in principles and personalities, Lucien Underwood judged the issue over the Rochester Rules largely in terms of professional and institutional attachments, and he classified the protestors into three groups. First were persons "connected with institutions whose *policy* is to retain the system in vogue at Kew and followed at Cambridge [Harvard];" second were those who so greatly admired Asa Gray that they saw the entire reform movement as an attempt to discredit him personally; and third were botanists who kept to themselves, never attended professional meetings, and "who know nothing of the spirit and aims of the recent movements in any personal way." Underwood concluded that the reforms were a necessary result of professional growth in America: "So long as all systematic botany was the *idea* of one man [Gray] such a system would work, but when as now there are a half dozen centers for systematic work in America instead of one there must be some system that is *uniform* and based on *principles* not chance; continued according to a *system* not by any notion that happens to be in the mind at the time."[55]

Thus as the controversy over the status of the Rochester Rules continued, and at times became quite heated, the reformers remained in control. As evidenced by the ease of acceptance of the rules and the checklist, the reformers dominated the AAAS and the major journals. With Bessey in charge of botany in the *Naturalist* and then *Science* after 1896; Arthur, Barnes, and Coulter the *Gazette;* and Britton the *Bulletin of the Torrey Club,* and *Science* until replaced by Bessey; the reformers held the major means of communications. Their control was further enhanced with the publication in 1896 of the first volume of Britton and Brown's *An Illustrated Flora,* which used the Rochester Rules, and with the adoption of the rules in publications of the USDA. In addition, some local publications such as those of the state botanical surveys of Nebraska and Minnesota, headed by Bessey and Conway MacMillan, adopted the rules.[56] While the Checklist of Flowering Plants was intended to provide a practical test of the usability of the rules, the use of them in additional publications could not be defended in the same fashion. The authors in these publications were obviously proceeding to apply the rules without international sanction.

As the situation now stood, there were at least four distinct reform groups in Germany, England, and the United States and many offshoots of each one ignoring or attempting to modify the Paris Code. All were using some of their own ideas in publications, yet none had the sanction of an international congress. In light of this, those American reformers who favored immediate use of the Rochester Rules in publications argued that there really was no international code in effect and that Americans were left with no choice but to use their own code since it at least had national approval.[57] While the controversy over nomenclature threatened the unity that promoters of professionalism hoped for, to many the controversy highlighted the maturity of American botany by showing that Americans were ready to participate in, and possibly lead, a reform of international significance.

As the nomenclature controversy raged, promoters of professionalization moved to complete the organization of the Botanical Society of

America (BSA). At the 1893 meetings of the Botanical Club when the society had been created and the twenty-five charter members selected, William Trelease was chosen to head a committee to create a constitution for the 1894 meeting. In consulting the charter members the Trelease committee had received a positive to enthusiastic response from about one-half of those elected.[58] As with nomenclature reform, the greatest hesitancy and opposition to the organization came from the botanists at Harvard and Yale.

Daniel Eaton of Yale wrote Trelease that he opposed the organization for a number of reasons, chief of which was his opposition to most of the persons associated with the new society. The same persons aggressively pushing for the new society were active in support of the Rochester Rules, and Eaton feared the reformers would use the new society to gain acceptance of the rules. He indicated that "with all this nomenclature and priority business I have not the slightest sympathy and shall utterly refuse to be bound by any rules such as those approved of by the botanical section of the American Association."[59] In a second letter to Trelease in which he again attacked the Rochester Rules, Eaton concluded that "I can have no scientific fellowship with the men who are now trying to control American botany and botanists, and I must decline to be a member of any association, society, or congregation which they are conspirators."[60] Eaton also opposed the new society because it originated in the AAAS of which he had not been a member since 1863. He resented botanists of the AAAS dictating policies to nonmembers, and he saw the new society being tied to the AAAS, which was controlled increasingly by midwesterners who were frequently moving the annual meetings out of the East.[61]

The Harvard botanists registered their opposition to the proposed society but in a more restrained fashion than did Eaton. The key persons in the Harvard-Yale camp were Benjamin Robinson and William Farlow. Eaton mentioned that he would largely follow the lead of the Harvard and Johns Hopkins people and that he had discussed the proposed society with Farlow, who in turn had to consult with Robinson. Roland Thaxter of Harvard likewise indicated to Trelease that he had delayed his response until he had learned the sentiments of Farlow.[62]

Like Eaton, Farlow thought the proposed society was too closely associated with the AAAS, and the AAAS was too much under the control of the "nomenclature crowd" and the promoters of professionalization who wanted the Botanical Club to be the forum for deciding matters that affected all botanists, whether members of the AAAS or not. "I see what it is all coming to," Farlow stated. "Britton, Coulter and company are to run the societies and committees and they in turn are to describe what is to be done and who is to do it." More basically Farlow had a generally low opinion of most American botanists, and he dreaded having another journal and more papers that would be the "same old trash" and that would serve only to further display the second-rate nature of American botany.[63]

Despite their opposition Robinson brought his Harvard colleagues into the new organization. Robinson responded to Trelease that the "enterprise does not find much favor at Harvard either from Farlow, Sargent, or Thaxter," but he added that if the society came about he would support it.[64] Neither Roland Thaxter nor Charles S. Sargent indicated that their opposition was in any way connected with the nomenclature controversy; indeed Eaton accused Sargent of being among the reformers. The reasons they related were shared by many other charter members—that is, the high cost of membership, the belief that another society and proposed journal were not needed, and merely a general opposition to societies.[65]

The constitution formulated by the Trelease committee provided for admission only to researchers, with the added provision that there be no honorary members or "old-timers" admitted. Only current work of a standard judged acceptable by peers would qualify a person. To force continued activity, a member had to attend the annual meeting and present a paper at least once every three years in order to retain membership. Finally, to insure that "only the committed would join," a high fee of $25 was required for admission in addition to the annual assessment of $10. A provision in the original draft providing that the society would always meet at the time and place of the AAAS was deleted and this issue was left open.[66]

At the first official Botanical Society of America meeting in 1896, Trelease presided as the group's first president and Bessey was elected to that position for the next year. For his presidential address, Bessey presented his second major paper on classification, "Phylogeny and Taxonomy of the Angiosperms" (see Chapter 8).[67] In subsequent contributions to the society, Bessey presented papers in 1899 and 1903 and was elected to the council in 1900.[68]

Some of the doubts about the advisability of the new society were confirmed by the small attendance at the early meetings, and as long as the nomenclature controversy boiled it remained divisive. During its first five years the BSA, instead of growing in membership, declined to twenty-one persons. Attendance was between ten and fifteen and the number of papers read ranged from four to eleven.[69] Particularly in the early years of the BSA there was difficulty persuading nomenclature conservatives and those who had been unenthusiastic about the organization to serve as officers. Both John Donnell Smith of Johns Hopkins and Liberty Hyde Bailey of Cornell, for example, refused the treasury after being elected in 1895. Although Bailey offered the excuse of being too busy and planning to go overseas, he wrote Barnes that the reason was the Rochester Rules.[70] In trying to influence Bailey to take an active part in the BSA, Barnes stated that "I assured him that at least twelve of the present members desired only cautious nomenclature reform and at least five of them no change at all . . . I don't think the society will devote its time to the nomenclature business, at least it will not with my consent. We want to promote botanical knowledge not discussion."[71]

At the same time, Bessey wrote Barnes that the new society should proceed slowly in admitting new members as "we cannot afford to run the risk of filling up with those not in complete sympathy with the movement." It appeared here that Bessey was referring to the movement to professionalize, promote the "new botany," and maintain the BSA as a select organization, not to the nomenclature controversy.[72]

Disagreement did not end, however. Through the first five years nine persons—including W. R. Dudley of Stanford, William A. Setchell of California, John W. Dawson of McGill University, W. W. Rowless of Cornell, William A. Kellerman of Ohio State, Lester Frank Ward of the U. S. Geological Survey, and Farlow—had refused membership, and three—Frank Lamson-Scribner of the USDA, Bailey, and Thaxter—had resigned. In 1898 five more botanists indicated their desire not to belong. There was enough concern that nominated persons would refuse membership that Trelease suggested the nominator obtain willingness to join before an actual invitation was extended "to avoid the humiliation of having him decline."[73]

Certainly if the BSA were to be the highest honor attainable in American botany, as its promoters intended, something had to be done to counter the charges of partiality and exclusiveness and to remove any doubts that it existed for anything other than to promote botanical research. Assurance of impartiality by the BSA was not helped by cases such as that of Erwin F. Smith of the USDA, an outspoken critic of the Rochester Rules. By 1898 Smith still had not been nominated for membership because it was generally conceded by the committee that the members would vote him down as a result of the personal bitterness he had generated. Bessey, who was on the nominating committee, agreed that there was danger of the membership rejecting Smith, but he favored putting Smith forward anyway because all on the committee at least concurred that Smith's research entitled him to belong.[74]

The only other issues in the BSA that attracted much attention during the 1890s aside from nominating procedures and dues were maintaining a high level of scholarship in the papers; deciding whether papers should be read or summarized, the latter a method strongly advocated by Bessey; and determining whether the money of the society should support research in the form of a journal, occasional memoirs, medal or money prizes, or subsidizing laboratory tables at research institutions.[75]

In 1899 the constitution of the BSA was revised to allow a larger membership with the stipulation that there would now be three levels of membership—life, associate, and patron. This step was a departure from the original intent to maintain an elite composed only of proven researchers. The founders had intended to keep membership small and by having high initiation and membership fees to keep out those who would join only for the prestige of association. The idea of associate membership was to admit young botanists who qualified but to assess them lower fees. This change was dropped two years later when all new members started with associate status.[76]

Despite the difficulties encountered with the BSA, Bessey and the other promoters of professionalization now believed they had provided a botanical hierarchy with the Botanical Club of the AAAS at the social, business, and less-scholarly level, Section G of the AAAS at the middle level with scholarly but general botanical papers, and the BSA at the top level where the best researchers presented their best work.

All through the 1890s Bessey, Coulter, Barnes, and Arthur continued their evaluations of the botanical meetings, but by the end of the decade the tone of their references to the meetings changed. It now indicated a belief in having accomplished their goals of building a professional base within the AAAS and of no longer needing to convince botanists that the meetings were valuable. The Botanical Club had been challenged by some during the nomenclature controversy as being not representative of American botanists and therefore not a proper place to make decisions of such magnitude as the Rochester Rules and the Checklist of Flowering Plants, but in general this question seemed settled with the club recognized as more representative than any other body to act in the name of American botany.

The first serious threat to the institutional dominance of the AAAS came in 1895 when nine botanists, including Farlow, Robinson, and Thaxter, discussed the desirability of organizing an affiliate society in conjunction with the American Society of Naturalists (ASN).[77] Until this time few botanists had been associated with the ASN. Organized in 1883 strictly as a northeastern meeting, during the 1890s the ASN attracted an ever-larger following among zoologists and some geologists, paleontologists, and psychologists, particularly as the AAAS met more frequently away from the eastern seaboard. Also, more specialized groups attracted members away from the general sections of the AAAS.

Although recognizing that a new organization would be primarily attractive to eastern botanists, a committee of five persons, including Bessey as a representative of western and AAAS botanists, tried to determine the interest and purpose of the proposed society. The advocates denied that there was any "antagonism to the existing organizations" or that there was any reason other than the desire to have a winter rather than a summer meeting. Bessey reported that he found little favorable response to the proposed organization among westerners as distance prevented attendance at two sets of meetings a year, and, at least in Nebraska, the Christmas vacation period was when the university hosted the state organizations such as the State Horticultural Society and the Academy of Sciences. The different seasons of the meetings would remain a factor only until 1902 when the AAAS moved to the Christmas period.[78]

Determining that there was insufficient interest in another general botanical society and that a specialized organization was needed, the organizers formed the Society for Plant Morphology and Physiology and held their first meeting in December 1897 in conjunction with the ASN. Bessey attended the 1898 meeting and presented a paper, but he never became very active in the group.[79] The nature of the new organization gave it wide appeal in that only its affiliation with the ASN made it an eastern meeting, yet it met the demand of many of the new botanists for further institutional specialization. Thus in part it was an expression of opposition to those who controlled the AAAS botanical structure and in part it was not.

As before, Bessey did not confine his entire promotion of professionalization to the AAAS. His main professional duty continued to be the editorship of the botanical section of the *American Naturalist*. Surviving the financial crisis of the early 1890s, the journal found the competition still formidable, especially with the revival of *Science* in 1894. After a decline in the size of his section in the early 1890s because of his acting chancellorship at the University of Nebraska, Bessey brought the number of notes in the section back to its earlier level. Edward D. Cope, the general editor, continually prodded Bessey for more botanical material during 1893–94, and his dissatisfaction became more evident. Finally in early 1895 Cope informed Bessey that he was dividing the botany department. Cope wrote that "I have written to you occasionally as to the pushing of the botanical department of the *Naturalist*." Cope continued, "several times recently botanists have complained to me of its restricted proportions . . . the quality you give us is just what is wanted but it is clear that you are too busy a man to give us quantity."[80]

Erwin F. Smith was one of those persons critical of the coverage in the *Naturalist,* and he and Cope had agreed that he would be the other botanical editor. At first Cope proposed to divide the sections into flower and nonflowering plants in which Bessey would have his choice of sections. When both Smith and Bessey agreed that such a division was not the best, Bessey suggested that Smith should handle physiology and pathology and he would do the remainder. Although Smith opposed this division, this was the arrangement Cope settled upon.[81]

Despite Bessey's concern with physiology and pathology as the nucleus of the "new botany" and his own continuing interest in these areas, he had always leaned more to structural rather than chemical physiology, and his interests during the 1890s were turning more to classification and plant geography. In the *Naturalist* he certainly had not given as much attention to physiology and pathology as the new specialists desired and deserved.

Yet Bessey suspected there was more behind Smith's dissatisfaction with the botany in the *Naturalist* than its slighting of pathology. In a privately published attack on the Checklist of Flowering Plants and in a letter to the English *Journal of Botany,* Smith charged the *Botanical Gazette* and *Science* with suppressing his writings in opposition to the Rochester Rules, and obviously Bessey had not presented Smith's side in the *Naturalist.*[82] Smith used his first opportunity as editor of the "Vegetable Physiology" section of the *Naturalist* to again attack the reformers. This drew a response from the *Gazette* that because Smith's attack had nothing to do with physiology it should have at least appeared under the general botany section of Bessey's.[83]

This uneasy situation of a divided botany section continued until 1897 when the death of Cope and an invitation to head the botany department of *Science* ended Bessey's seventeen-year association with the *American Naturalist.* Under Nathaniel Britton, the current departmental editor at *Science,* there had been very little botany in the journal and never a section of notes as Bessey had published in the *Naturalist.* J. McKeen Cattell, the general editor, asked Bessey to enlarge the section and to duplicate what he had done with the *Naturalist,* which the Nebraskan quickly accomplished.[84]

Bessey's association with *Science* was to be long and harmonious. Britton, who was now associated with the New York Botanical Garden, remained on the editorial board. Although he occasionally threatened to resign Bessey always persuaded him and Cattell to maintain the status quo. Bessey and Britton were very alike on professional issues. More importantly, Britton generally had little to do with the active editing of the botany section, leaving the decisions concerning content to Bessey, an arrangement Bessey liked. Bessey's control was somewhat threatened after 1900 when Cattell apparently was under some pressure, not to replace Bessey but to include a broader representation of views on the editorial board. Although Cattell told Bessey that the botanical section had no problems and was one of the most active, the editor decided to reorganize all the sections in 1904. Part of this change was mandated by the AAAS, which now required the vice president of each section to be on the editorial committee, and in botany the new committee added John M. Coulter and William Trelease to the incumbents, Bessey and Britton. Despite the changes Bessey retained final authority for botany content, and he was able to continue "certain general lines of policy which I should wish to control" until his death in 1915.[85]

In an area closely related to promoting botany, Bessey continued his assistance in developing an institutional structure for agricultural science. During the 1890s Bessey and other botanists continued their participation in the Society for the Promotion of Agricultural Science (SPAS). The problem of the SPAS from the botanists' viewpoint was not their activity but rather the overall lack of participation, the sluggishness of the society, and the lack of outside respect.

Only slowly did agricultural science gain recognition in the late nineteenth century. In addition to the growth in numbers and specialization, agricultural science gained most during this period from the maturing of agricultural colleges and the escalating prestige of the USDA. Symbolic of the gains being made in the acceptance of science by the agriculturists was the consultation of scientists by the agricultural press and the initiation of regular sections concerned with science. For example, the *American Agriculturist* started its section "Science in Farming" in 1889. Usually this section contained summaries of agricultural bulletins issued by the USDA and the experiment stations and a profile of an agricultural scientist. The biographical sketch about Bessey appeared in July 1891. The section also gave special attention to the proceedings of the SPAS.

Only by making agriculture scientific could the new professionals feel at ease and have pride in their line of research. Thus agricultural scientists worked to duplicate the professional standards that they were bringing into their various scientific specialties. Essential in this regard were higher academic standards for agricultural classes and more scientific competency in the USDA and the state experiment stations. Certainly the members of the SPAS intended to provide the leadership and the model to bring about these changes.[86]

The achievement of these aspirations remained in doubt throughout the 1890s, however. Within the SPAS, despite increasing the membership limit to one hundred, attendance at annual meetings remained small, usually under twenty. Likewise the number of papers declined, and some members believed the quality of the papers lessened. A few members continually pleaded for more circulation of research results through a new journal. On this issue Bessey agreed that a weakness of the SPAS was that its work was not widely enough circulated, but he opposed the society's taking on the financial and editorial responsibility of its own journal. When there was discussion of taking over the *Journal of Agricultural Sciences,* a private journal, Bessey argued that the journal already was good and would probably only be hurt by being associated with any society. Most importantly, he believed the SPAS was having enough trouble merely remaining in existence without assuming the burdens of a journal. In fact, in the mid-1890s Bessey stated that he "seriously doubted the advisability of continuing" the SPAS, but he was won over by those persons who believed that there was no other organization that could fill the niche of the group and that its extinction "would be a loss to science."[87]

In contrast to the SPAS, the Association of American Agricultural Colleges and Experiment Stations (AAACES) prospered during the nineties. Nevertheless, the interest of agricultural botanists and scientists in general remained minimal. Having persuaded the AAACES to organize scientific sections, the scientists seldom attended enough to justify the sections' existence. As for the section of botany and horticulture, a special effort in 1896 resulted in scheduling one horticultural and four botany papers, but there was only one additional botanical paper during the next two meetings.[88]

Despite the problems of the SPAS, some members were optimistic that their work was "fast gaining favor with college and station investigators and with all who are interested in the advancement of the science of Agriculture."[89] By 1900 enough members, including Bessey, had a renewed hope and interest in the SPAS by pledging attendance and papers to ensure that the organization would remain viable.[90]

A major shift in organizational orientation also seemed temporarily to solve some of the group's problems. In the early days members of the SPAS had seen their natural ties with other scientists in the AAAS and had arranged their meetings to coincide. That way members such as Bessey could arrive one day early to attend the SPAS meetings and then go on about their regular scientific interests in the AAAS. After the turn of the century membership changes and shifts in attitudes as to where the society rested in the overall scheme of scientific endeavors led the SPAS to gravitate toward such groups as the Society of Horticultural Science and the American Society of Agronomy. Recognizing that its membership now more often attended meetings of the AAACES, the SPAS in 1906 shifted its meetings to coincide with that group rather than the AAAS. In 1911 the SPAS joined the Affiliated Societies of Agricultural Science. Specialization had occurred to the point that agricultural botanists, for example, felt more at home with other agricultural scientists than with other botanists. This shift obviously did not please Bessey.

The SPAS continued to struggle throughout the early 1900s. Each organization in the Affiliated Societies retained its autonomy, and efforts to embrace all agricultural scientists within the SPAS failed. Also, the AAACES ceased to organize around scientific disciplines and became primarily concerned with administrative issues. Efforts to find a proper home for the SPAS drifted until attempts to draw the agricultural scientists back to the AAAS resulted in the forming of a new section for agriculture in 1919. During the next few years the various agricultural science societies gravitated to the new section in the AAAS, and the SPAS merged with the AAAS in 1923. The agricultural scientists were now united with other scientists, as Bessey had always wanted, but they joined only with the recognition of their particular interests and with their own section.[91]

Despite the developments in agricultural science, many scientists continued to regard farm-related studies as second-class, and some were even hostile to them. The late nineteenth century was a time when many scientists in all disciplines were attempting to awaken Americans to the reality of their dependency on Europe, and many believed that the major drawback to progress in the United States was the constant stress on application. John Tyndall, the visiting English scientist, observed that applications of science dominated American science rather than "science for its own sake, for the pure love of truth." Simon Newcomb and Henry A. Rowland, leaders in American astronomy and physics, made loud and frequent comments along the same line. Rowland called for "disinterested" scientific research. Both agreed that what was holding back American science was the confusion between pure science and its applications and that this confusion was hindering all major intellectual achievements. They saw the growing interest in agricultural science and the growth of the USDA as prime examples of American stress on application and the further misdirection of American energies and resources in science.[92]

Bessey agreed completely with Newcomb, Rowland, and other purists that American scientists needed to do more original research. He did not, however, make the same distinction between pure and applied research. Bessey saw these as two divisions of science but did not believe one was inferior to the other; one division was not necessarily scientific and the other division nonscientific. The principles of biology, chemistry, and physics had to be based on experimentation to be scientific, Bessey believed, and for agriculture to be scientific it, too, must use experimental procedures.[93]

Here Bessey was confronting a perplexing and ever-present problem faced by all applied disciplines. Bessey's farm background, his schooling, and the influences of his academic and administrative career at land-grant colleges blended with his training and associations in basic science to blur most of the avowed distinctions between science and technology. It was this middle tradition that allowed scientists such as Bessey to share some of the perspective of both purist and practitioner and to feel at ease in public science. Certainly the land-grant movement, as conceived by Bessey, would provide the ideal intellectual environment for the combination of pure and applied science. Bessey fought, and eventually would lose in his own university, to keep the new agricultural schools within the regular university structure, and he intended for agricultural scientists to remain largely within their own scientific disciplines. Instead the agricultural scientists of the twentieth century soon had their own organizations, standards, and sets of rewards.

While there was sincerity in the attachment to this middle tradition, this perspective provided agricultural scientists, engineers, and government scientists in general with a rationale for their professional status. The leaders of the agricultural science movement, to be successful, had to adopt professional standards and scientific methodology to win acceptance within the scientific community, and they had to convince the lay community that scientific standards were the proper ones upon which to judge the work of academic and governmental agriculture. Success required sharing the values of a tradition that recognized pure science but that did not make a great distinction between science and technology.[94]

Criticism of agricultural science by both extremes—purist and practitioner—thus bothered Bessey. In response to the purist Bessey noted little difference between pure and applied research. To Bessey it was the method that determined whether work was scientific or not rather than whether the experimenter intended to apply the findings directly and immediately. Was Liberty Hyde Bailey any less a scientist because he applied his botanical knowledge to horticulture, or was Erwin F. Smith any less a scientist because he worked for the USDA on plant diseases? Bessey believed not. He agreed with W. H. Jordan of the New York Agricultural Experiment Station that the United States needed universities "where the flavors of the farm do not damn a subject to intellectual perdition."[95] The growing number of agricultural scientists, Bessey thought, were making contributions to scientific knowledge because they were good scientists. The agricultural scientists pointed to the French scientist Louis Pasteur to show how practical research, if well done, could greatly contribute to scientific knowledge. By the 1890s similar work was being done by American agricultural scientists. Admittedly, noted Bessey, there were too many persons working in agriculture, even at the experiment stations, who were not scientists. However, they were not scientists not simply because they were working in agriculture but rather because they did not have an adequate grasp of the scientific principles upon which agriculture must draw and they did not know how to develop principles and solve problems by experimental methods.[96]

Speaking to the purists, Bessey exhorted botanists to recognize the potential good of the agricultural experiment stations. "It will not do for us to stand aloof, and decry their results as not accurate, and as agricultural instead of botanical," Bessey stated. Instead, botanists needed to work more closely with the experiment stations. Such cooperation would improve agricultural science; likewise, cooperation would awaken botanists to the contributions that were possible from the stations. The result, Bessey hoped, would be to develop the agricultural experiment stations into "centers of investigation of the greatest importance to science."[97] Bessey further cautioned the purists that "no science can hope for support or recognition that does not respond to the demands of its age." He certainly did not want to detract from pure research or from the prestige of the "prophets who prepare the way for the oncoming of scientific truth," but he believed that "it is only when such truth has permeated contemporary society that science survives."[98]

Although he was much closer to the purist camp than to the applied, Bessey wanted to draw the two areas closer together by making both experimental. In his writings he devoted less attention to the purists. Most of their objections he saw as snobbishness and an exaggerated feeling of importance for their own area of specialty. The practical agriculturists, he believed, were the greater obstacle to agricultural science. He disagreed with those persons who wanted the USDA, the farm societies, and the agricultural colleges to do only practical work and who believed that the study of basic botanical principles was a waste of time for the farmer or fruit grower. Therefore, his efforts in writing to and addressing boards of regents, farm journals, newspapers, farm institutes, and educational groups were to emphasize the good that science could do for agriculture.[99]

With the turn of the century, the disruptive features accompanying the growth of American botany remained, but a mood of conciliation was more frequently in evidence. The combatants in the debate over the continued use of the Rochester Rules had calmed in anticipation that the approaching International Botanical Congress in Paris in 1900 would provide the needed authority to impose a settlement. Yet there was very little discussion of the congress in American journals, and the Paris congress, on the subject of nomenclature, proved no more capable of finding an acceptable solution than had previous attempts. The congress "declared itself incompetent to revise the laws of nomenclature" and merely tried to provide a basis for a better effort at the next congress in 1905 by establishing a committee to coordinate preparations.[100]

Between 1900 and 1905 the discussion of the Rochester Rules continued but at a much-diminished level. The points of contention had been made and remade, and little more could be done to cause a shift in positions until an international meeting could accomplish a revision in rules that would receive enough support to overcome the national authorities. Merritt L. Fernald of the Gray Herbarium was the chief public critic of the Rochester Rules, but more and more Americans were adopting all or part of the reforms. Yet among most reformers there was always an uneasiness that their ideas did not have international sanction, and this feeling increasingly strengthened the position of moderates who saw that a compromise code was the best that Americans could hope for unless they were willing to follow the radical reformers and the English and continue to pursue a strictly national path.

Aside from the occasional article or letter that appeared in the journals, the main reminder of the differences of opinion over the rules was the often barbed remarks in book reviews. The nomenclature committee of the Botanical Club remained in existence under the leadership of Nathaniel Britton, and it continued to assemble a code, hoping to accommodate as many Americans as possible and to submit its proposals to the congress in 1905. The committee reported to the Botanical Club in 1903 that the code was ready for a final consideration and recommended a special commission for that purpose. In a special meeting of the nomenclature commission in Philadelphia in early 1904, a proposed Code of Botanical Nomenclature was approved and subsequently submitted to the international congress.[101]

The Vienna International Botanical Congress in 1905 finally accomplished what previous ones had not — an official revision of the 1867

International Code of Botanical Nomenclature. For Americans it did not end the controversy over the American code, but it certainly put the issue in a different perspective.[102] As Charles Barnes concluded, "On the whole the action of the congress was conservative from the American point of view, yet it marks great progress toward a stable nomenclature. . . . When the final editing is accomplished and the new code is available, we may reasonably expect the rules to be generally followed until a further revision is possible."[103] The response of Barnes, a moderate, was not shared by more fervent reformers, and the congress was extremely disappointing to most members of the nomenclature committee and commission.[104]

Several more decades passed with American systematists divided, some adhering to the American code and some to the international code. The divisiveness of the issue had less impact with time, however, and nomenclature captivated the interest of fewer and fewer botanists, even systematists, as the impact of the "new botany" continued to move attention of the discipline toward more important issues.[105]

Aside from nomenclature at the turn of the century, the schism over the proper institutional home for botanists and the uncertainty regarding how to accommodate continuing specialization remained serious threats to national unity and to the achievement of basic professional gains envisioned by promoters of the "new botany." The same impulses that had motivated biologists to separate from natural scientists in 1880 and botanists from zoologists in 1890 now threatened to disrupt the recently established institutional structure within the AAAS. In addition, the growth of botany across the nation posed serious problems of sectional pride, and distance alone made attendance at national meetings increasingly difficult. Was it to be impossible for any society to represent all botanical specialties, all views, and all geographical regions?

The promoters of professionalization made a special effort in 1903 to use the patriotic appeal of holding joint meetings of the major scientific organizations in Washington, D.C., to bring the nation's botanists together. With the AAAS and the ASN meeting jointly, it meant that more than one hundred botanists who came to the nation's capital could participate, if members, in all the major societies. Even the Botanists of the Central States met in conjunction with the Washington meeting.[106] Disunity was a subject of considerable discussion in Washington and resulted in the Botanical Society of America and the Society for Plant Morphology and Physiology naming committees to exchange views on how to achieve "cooperation" with each other and, if possible, "federation." Although a number of persons were involved, the major proponents of closer ties between the societies were Bessey, William Ganong of

Smith College, Frederick Atkinson of Cornell, Charles Barnes of the University of Chicago, and Beverly T. Galloway of the USDA.[107]

Galloway made national unity and the problem of "extreme specialization" major themes in his presidential address to the Botanical Society of America. Lamenting that problems of specialization and the multiplication of societies could be solved only by botanists joining forces for the advancement of their discipline, Galloway concluded, "The time seems ripe for bringing about this result. Never was botany more prosperous, never more aggressive. On the threshold of the twentieth century we stand, knowing our strength and only needing to weld it into harmonious action to make it vital and lasting. Let us join hands and do our best to bring this about."[108]

Although preliminary steps were taken for cooperation, an actual merger seemed extremely remote at first, yet further divisions in the form of new societies made the issue of merger increasingly significant. Additional splintering had occurred in 1899 with the creation of the Society of American Bacteriologists and in 1903 with the establishment of the Botanists of the Central States and the Society for Horticultural Science, followed in 1904 with the American Mycological Society.

The committee for cooperation of the Botanical Society of America and the Society for Plant Morphology and Physiology was joined by representatives from the new American Mycological Society, and the enlarged committee proposed a plan of merger during 1904. None of the groups involved wanted to lose its own identity, and the nature of the groups was not outwardly compatible. The Botanical Society of America identified itself as a general botanical society, representing the elite from all areas of botany, yet it was the society most troubled by inner dissension, resignations, and outside criticism. The Society for Plant Morphology and Physiology and the American Mycological Society, on the other hand, were specialized and less select in membership requirements and fees. Generally the Botanical Society of America believed the other groups would merge with it, use its name, and allow only the best of the members of the other two societies into the merged organization. At issue therefore were a number of related questions. Should there be different levels of membership to recognize the different degrees of accomplishment? Should the dues be high or low? Should the basic philosophy be one of exclusiveness or democracy? Where and when should the meetings be held?[109]

The wave of sentiment, evident in the past several years, to subdue intradisciplinary, sectional, and personal squabbles for the sake of greater professional gain was evident in the willingness of each society to compromise. The three societies agreed upon a formal merger in December 1905. The merger retained the Botanical Society of America as the name, established two classes of membership with the distinction based on published research, set fees at five dollars (which was a compromise), and divided the paper sessions at the meetings into specialized sections. The first meeting was held in conjunction with the AAAS, but the new society left flexible the time and place of the meetings.[110]

Most leaders in the professionalization movement believed unity was essential, yet specialization was going to continue, and it had to be accommodated. Also, experimental botany, with its increasing concern for studies related to embryology, heredity, and evolution, was involved in subjects that pulled botanists into associations largely dominated by zoology and general biology.[111] Enough botanists had agreed that a single society was preferable for professional and social purposes to achieve a merger, but many recognized that it was not a final solution and that organizations would continue to subdivide along lines of specialized research interests. William Ganong conceded that "I think it a great advantage for me, for instance, to sit around the same beer-table with workers in Algae and Fungi, about which I know mighty little, and to hear an occasional seminary lecture by an expert upon these subjects." But he exposed the dilemma by stating that "I do not see what advantages there are to me in attending their meetings and listening to their awful dissensions, when I might be with you [Charles Barnes] and MacDougal and True and others hearing something I have some interest in."[112] The Botany Section of the AAAS had finally to succumb to the pressures of overcrowdedness and specialization, and it divided into subsections in 1909.

The growth in size of the profession and the changing nature of the subject matter necessitated the institutional changes that had occurred since 1892, but the advantages and disadvantages of alterations were not always clear and adjustments not easily made. Through all these problems, and others, associated with growth and professionalization, the objectives of the new botanists of the 1880s had largely been fulfilled. Even the goal of putting American work on a level with that of Europe, as highlighted in the nomenclature controversy, had in part been achieved, and in their most optimistic moments Bessey and the other American promoters could boast, with national pride, of the great distance American botany had traveled in the past twenty-five years. Pioneers like Bessey had built a good foundation, and in the early 1900s they began to pass active leadership to the persons they had trained who would guide the next level of professional growth.

# 6

# Grasslands and Forests

FROM THE TIME of his 1884 arrival in Nebraska, one of Bessey's major projects was to have an accurate and extensive knowledge of the plants of the state. Immediate practical concerns of the State Board of Agriculture and later the Agricultural Experiment Station centered his attention on the grasses and forage plants of Nebraska, but his long-range botanical interest remained the entire flora of the state. What knowledge there was at the time was confined mainly to the eastern third of the state and even in this area the information was uneven. Samuel Aughey, then a professor of natural history at the University of Nebraska, had published a catalog of Nebraska plants in 1875 but by standards of even a decade later this was judged "sadly defective." The only other early attempt to systematize the plants of the region was the work of the Hayden Survey in the early 1870s.[1] As one of Bessey's students remarked in 1889 regarding the Nebraska Sandhills, it is "not particularly interesting from its rare or remarkable flora, perhaps but from the general ignorance in regard to it."[2]

As with his early state survey in Iowa, Bessey wanted to know what plants grew in the state and where. He could then classify the flora and fit this information into the larger national picture. While most such projects at the time resulted in lists of plants, Bessey viewed distribution as increasingly important. Distribution could provide answers regarding the history of the flora within the region, what plants were moving in and out, the origins of the local floras, and changes in and interrelations with the environment. There were also related interests in a state survey that were confined to the internal working of systematic botany. Bessey in particular had been developing ideas regarding the natural sequence of plants and how this sequence should be presented in published floras. He wanted to cover all plant groups and not favor the flowering ones, and he wanted to arrange the flowering plants differently than was usually done in earlier plant surveys. Bessey desired more than a mere listing of plants to result from the Nebraska survey.

With little to guide him in printed sources, Bessey had to learn most of the Nebraska flora firsthand. By the late 1880s his field trips began to take him beyond the vicinity of Lincoln, and he tried to take advantage of his increasing speaking obligations at Farmers' Institutes, state agricultural meetings, and commencement programs to see more of the state. Whenever possible he arranged these trips with time for at least a short excursion into the countryside to observe or collect specimens. Especially helpful was the 1887 completion of the railroad across central and northwest Nebraska to the gold fields of the Black Hills in the Dakota Territory. This offered convenient accessibility to a vast new botanical area, which Bessey soon explored.[3]

With his increasingly heavy teaching and administrative load in the late eighties and early nineties, Bessey relied on students to help in his plans for a state botanical survey. He began to cultivate a small but steady number of intelligent, dedicated, and highly motivated botany majors who upon graduation stayed at the university to form the basis for a graduate program. In 1886 seven undergraduate science students banded together to defend themselves from the taunts of the classical students—the "Lits"—and to pursue their enjoyment of field botany. Five of this group (Roscoe Pound, Thomas A. Williams, Herbert J. Webber, Jared Smith, and Albert F. Woods) who continued through to graduate study provided the nucleus for the organization, which by 1888 was more formally organized and termed merely the "Sem Bot." Also in 1888 the departments of botany and history organized German-style seminars devoted to research. Originally a secret body, the Sem Bot by 1891 admitted its first new member, opened its formerly private paper-reading meetings, and provided botany students with a type of honorary society and the basis for an academic esprit de corps. Membership in the Sem Bot remained very limited during the first decade; an examining committee tested prospective members, admitting only seven from the first fourteen candidates. Classes of membership, some secret fraternal ritual and mostly good-natured ceremony, caps and sashes, and eating of mincemeat pie became proud traditions of the group.

Central to the nature of the Sem Bot were the give-and-take discussions and the pursuit of key and controversial issues in botany that reflected Bessey's approach to learning. Bessey encouraged the "eating and breathing of botany" with his open laboratory where students worked largely on their own and had free access to the laboratory, herbarium, and library six days a week (and occasional nights), and it was this atmosphere with Bessey at its center that produced such an outpouring of young professional botanists.[4]

Out of their informal field trips during the mid-1880s some order and pattern of collecting plants developed. Following the lead of Bessey, Webber and Smith started gathering field notes on the grasses of the state. Williams followed by making a preliminary investigation of the lichens and Webber the algae. The students also started regional studies of the state with Webber exploring the Sandhills of north-central Nebraska and Williams and Pier A. Rydberg, a new addition to the group, investigating the canyons and high plains flora in the northwest.[5]

What came from this early work by Bessey and his students was an incomplete but improved knowledge of the kinds and locations of plants in the state. By 1889 Webber and Bessey had compiled the first catalog of Nebraska flora, and in 1892 they added appendixes containing descriptions of over four hundred plants.[6] As they found a number of new locations they began to patch together the general pattern of the state's plant geography.[7] What was beginning to emerge was the picture of Nebraska as a transitional region. This aspect was particularly noticeable in repeated trips across the canyon flora of the upper Niobrara River that stretched across northern Nebraska. In one particular canyon of Long Pine Creek Bessey found both the ponderosa pine, which grew only westward to its established location in the Rockies, and the black walnut, which extended eastward to its established stands in the upper Midwest. As Bessey conjectured, "I doubt whether there is any other place on the continent where the Black Walnut and the Rocky Mountain Pine grow normally side by side."[8]

Following these early foundations, in June 1892 the Sem Bot organized the Botanical Survey of Nebraska to formalize and facilitate publishing their work. Of the early participants, Webber had graduated and moved on for additional study at the Shaw Gardens in St. Louis before gaining employment with the Division of Vegetable Pathology of the USDA. Woods followed Webber in 1894 to work for the USDA as did Smith and Williams shortly after. So complete was their attachment to the ideals of the Sem Bot that Woods, Smith, and Williams started a similar seminar group in Washington, D.C.[9] The students remaining after Webber's graduation (Smith, Woods, Pound, Rydberg, Frederic Clements, and Herbert Marsland) continued to traverse the state during their summer excursions with the major concentration continuing to be the Sandhills. The early years of the survey raised the number of recorded species for Nebraska from approximately 2,500 in the early nineties to 3,400 by 1897.[10]

The Botanical Survey of Nebraska as an activity of the Sem Bot had no official sanction, which both helped and hindered its progress. The Sem Bot, as Bessey described it, was not a part of the botany department at the university but an "ally" of it. By not having official status the Sem Bot had complete latitude to set standards for its work, and Bessey believed this was the strong point of the Botanical Survey. Following Bessey's lead during the 1890s, with Roscoe Pound and Frederic Clements providing the primary stability and permanency, the small group established rules for a strictly "scientific" survey. For example, they could not record any species unless they deposited an actual specimen in the herbarium for the survey, and they followed "progressive" or radical standards for nomenclature and classification. They adhered to the Rochester Rules, and they used Bessey's own revised classification system except when working for the USDA, then they followed its use of de Candolle and Bentham and Hooker.[11]

The main drawback to the private status of the survey was that it had no regular financial support. Bessey and the students volunteered their time, and attempts to obtain financial backing met with only limited success despite the encouragement given the survey by the university and by some state agencies. The State Board of Agriculture printed Webber's "Catalogue"—and later the additions—in its annual report, and plant distribution maps and summaries of the survey appeared in publications of the Agricultural Experiment Station, the State Board of Agriculture, and the State Horticultural Society. Also, as he had done earlier, Bessey was able to convince the Division of Botany of the USDA to hire a few of his students as botanical collectors during the summer.

In 1891 and 1892, for example, Rydberg worked in the Sandhills for the government. He continued his work there and in 1893, the USDA hired Clements to collect in the Niobrara valley. For this work each student received one hundred dollars a month with the agreement that the USDA received the first set of mounted specimens, the University of Nebraska the second set, and the student the third.[12] Each spring for the next few years Bessey wrote Frederick Coville of the Division of Botany with the summer plans of the Botanical Survey of Nebraska and offered the usual partnership on collections. The USDA was satisfied with the arrangement because Bessey's students proved to be able and ardent collectors. In 1896 Frank Lamson-Scribner writing from the USDA stated that the collections by Rydberg and C. L. Shear "were on the whole the finest ever obtained by the Department."[13]

To further supplement the meager resources of the Sem Bot, the group established an exchange club whereby interested amateurs around the state could trade plant specimens. A few of these amateurs, such as the Nebraska minister J. M. Bates, became regular contributors to the survey.[14]

Hoping to publish some material each year, in 1894 the Botanical Survey published the first two parts of the *Flora of Nebraska*. The survey anticipated twenty-five parts in all. The *Botanical Gazette* noted that while the *Flora* was primarily of local interest "it is of decided general interest as it not only represents an effort unique in this country, but it deals with one of those 'middle regions' that have never had fair treatment at the hands of the manuals." Parts I and II received glowing reviews in the *Gazette* and in *Science*. Both stressed the introduction by Bessey in which he presented his views on the major plant groupings. Nathaniel Britton, one of the reviewers, observed that Bessey's ideas on classification were certain to be controversial, and he concluded that these volumes "must long hold first place in the published results of the exploration and study of a local flora." The reviewer in the *Gazette* emphasized that the algae and fungi were to be included in the *Flora of Nebraska;* "that it begins with the lower plants or even treats them at all will be a revelation to many a teacher who thinks of these as plants to be sure, but hardly as plants which can be described, much less identified, by ordinary mortals."[15]

Having anticipated issuing three or four parts of the *Flora* each year until completion, the Sem Bot was unable to come close to these plans. The group issued only one part in 1895 and lacked money for further work until the university regents provided a grant of $500 in 1899.[16] Again, the reviewer of Rydberg's volume praised the attempt of the survey to demonstrate the evolutionary relationships of the various plant groups as well as the "careful" use of the new reforms which "blazes the way for those unfamiliar with the nomenclature."[17]

Thus by the mid-1890s Bessey and the Sem Bot had made a good start in covering the state, and in some locales like the Sandhills, their work was nearly completed. Although the first step of the survey was to record all the plant species within the state, Bessey conceived that function as only a start to a complete "scientific" survey. Much more significant, he believed, was the contribution the survey could make to understanding the state from the basis of plant geography and ecology. Bessey also believed that the survey could not restrict itself to state boundaries because they needed information on surrounding areas that were associated with or were extensions of the flora of Nebraska.

By the mid-nineties, therefore, members of the Sem Bot, as they had on a lesser scale before, began venturing into surrounding states on their summer collecting trips. Bessey, Williams, and Rydberg had all investigated the Black Hills; Rydberg, Shear, and Clements surveyed areas in the Rocky Mountains; Shear worked on Kansas flora as did Williams and DeAlton Saunders on South Dakota; and Bessey visited the Yellowstone region.[18] Increasingly the purpose of these trips, such as Bessey's to the high plains of western Nebraska, was to observe the geographical features of the flora as well as to collect specimens. To obtain an even broader view of the Great Plains, members of the survey studied the findings of the Canadian Geological Survey and the old Hayden Survey.[19]

In Bessey's initial work on the Nebraska flora in the 1880s, he had stressed that the significance of the region was its transitional nature. This feature continued to be the prevailing theme in the papers and publications coming from members of the survey in the nineties, and they illuminated the topic by providing further and more detailed examples. Williams in his study of the Black Hills compared the region's lichens with those in eastern Nebraska and those of the Yellowstone region as reported in the Hayden Survey. He concluded that the plants of the Black Hills were subalpine and related more to those species of the Rockies than those of the prairies of eastern Nebraska. Rydberg, in a paper on the "Floral Features of Western Nebraska," stressed that contrary to popular belief the prairies did not extend from Illinois to the Rocky Mountains. Instead, between the prairies and mountains was a distinct region, the Great Plains.[20]

While the distinctiveness of the Great Plains was certainly not an original idea with Bessey or his students, the concept of the Great Plains as a botanical region attracted an increasing amount of attention. More than any other group at the time the Nebraskans provided the information and conceptual framework for an understanding of the region and an appreciation of its uniqueness. Early in his study of the region Bessey had demonstrated that the predominance of movement of plants into the state came from the southeast, extending up the river valleys, and from the northwest, extending from the mountains.[21]

From the beginning of his interest in a state survey of Nebraska in the 1880s, a key goal for Bessey and his students was to understand the state's most prominent feature—the grassland—and the feature most conspicuous by its absence—the forest. Although most of Bessey's basic study of the native and introduced grasses and other forage plants of Nebraska had already been done by the late eighties and early nineties, he continued to provide progress reports on the accumulating knowledge of the region.[22]

Because some of these studies were intended for an agricultural audience, one of the obvious interests of the survey was the effect of cultivation upon the native flora of the region. These practical studies also were in line with the work of the survey as Bessey described the distribution, movement, and physical and climatic conditions under which grasses, forage plants, and weeds grew. Members of the survey gathered such information for scientific purposes, but it also reflected an increasing attempt by Bessey and the survey to use the ideas of plant geography and ecology to further their understanding of grasslands.

*Botanical Gazette* editor John M. Coulter, speaking of the increased influence of plant geography, stated that the nature of plant collection was certain to change—that biological surveys were replacing the mere collection of plants. Roscoe Pound picked up on this theme to further explain the change: "Geographical botany as it is now understood is comparatively a recent development." He continued that "collectors and cataloguers for a long time have been gathering a portion of the bare facts upon which geographical botany must proceed, and the facts of plant-distribution have been more or less ascertained. But the systematic collating and grouping of these facts . . . is a matter of the last few years and is still going on."[23]

This was the type of work the Nebraska survey had conducted nearly since its commencement. From recording the appearance of different species, usually by a political unit such as a county, and transferring this information to shaded maps, the growing influence of plant geography had moved the botanists to determine not just the appearance of a species in a given locality but also the numbers and conditions of an appearance. As Pound explained, "but the statistics as to the distribution of families . . . gave no promise of leading to important results. . . . It is apparent that a mere statement of the number of species of the various natural plant-groups occurring in a certain region tells us very little of the vegetation of that region except in the most general way." The objectives of modern plant surveys needed to go beyond this. Again Pound explained that to "understand the vegetation of a region one must ascertain not only what are its physical, meteorological and geological features, but much more what sorts of plants control its water, meadow, plain, or forest vegetation."[24]

The immediate source of inspiration for this expanded geographical or ecological approach to botany, at least for the members of the Nebraska survey, was Oscar Drude's *Deutschlands Pflanzengeographie,* published in 1895, in which Drude described the botany of Germany in terms of "vegetation-groups." To Pound the study by Drude marked a new era in plant geography. Unlike most earlier intensive geographical studies, Drude covered a large area, and he provided a model of how to classify a large, diversified region. The model contained both the methods and the theory upon which the classification of the "vegetation-groups" was based. Pound was convinced that one could apply Drude's model to other regions without additional methods or theory. "Vegetation-groups" were those aggregations of plants that exerted a "positive role in the vegetation of the region in question," and it was not the number of species that was of major concern but the "biological plant-community of the region." Pound contended that most of the newness of Drude's study was not only the general applicability of its ideas and methods but also that the "work is in some sort a summary of geographical botany as it now stands." Pound concluded that "so much material necessarily takes on a new aspect when brought together and digested, though we have been more or less acquainted with a large part of it in its scattered condition. As part of a whole, each fact seems something new."[25]

The first significant result of the Drude model for the Nebraska survey was the *Phytogeography of Nebraska* by Frederic Clements and Roscoe Pound and a follow-up article by the same two authors in which they expanded their study beyond the boundaries of Nebraska and attempted to classify the vegetation regions of the entire "prairie province." Following earlier attempts by A. H. R. Grisebach, Adolf Engler, and Drude to classify the vegetation of North America, Pound and Clements identified the inaccuracies of these studies and provided their own classification. Although Drude had divided the prairies into the "northern forest-prairie region" of Canada and the "Missouri prairie region," Pound and Clements made all prairie one province with subregions. Likewise Drude had a separate vegetation group in the strip along the eastern front of the Rocky Mountains, but the Nebraskans stated that "topographically and phytogeographically, the foothills . . . are an intrinsic portion of the Great Plains, and, as such, are to be included in the prairie province." In addition to prairie and foothill regions in the prairie province, they included a third region, the sandhill. All three regions were present in Nebraska and were particularly distinct so that the growing information gathered by members of the survey was valuable and provided the basis for clarifying the characteristics of the prairie.[26]

As members of the survey roamed over the three prairie regions and the adjacent wooded bluff, meadowland, and alpine regions, they increasingly studied the area from the perspective of these regional groupings and their subgroupings, or formations, as outlined by Pound and Clements. The region that Bessey paid particular attention to in the late 1890s was the foothill. In 1897 Bessey traversed this region in western Nebraska, starting from the rolling treeless grasslands at Alliance and heading west through this region to an area of sparse sandhills, then to the North Platte valley, on to the canyon region of Cheyenne Ridge with its pine-covered slopes, then to grasslands on top of the ridge, and finally to the piney Wildcat Mountains. Here Bessey saw and described a region where the vegetation had largely moved out of the mountains and the dominant factors operating were dryness and altitude.[27]

Aside from grasslands, trees were the other prominent feature studied by Bessey and the Nebraska survey. For a botanist who spent his entire adult life on the prairies, trees—or their absence—were of necessity a constant topic of interest and concern. Seldom did meetings of the state agricultural or horticultural societies or the academies of sciences pass without debating the consequences of trees. Just as links between rainfall and the plow were pondered, so too the question of whether rainfall followed the tree was a subject of considerable speculation. Could trees moderate the temperature, slow down the winds and thereby reduce evaporation, or increase humidity through respiration? A closely related question to the relationship between rainfall and trees was the history of the prairies. What caused prairies and plains to be largely treeless? Had they always been this way or was the treeless condition merely a stage that could be changed with climatic or human modification?[28]

Even though science could not yet provide conclusive answers to these questions, the general opinion in the late nineteenth century was that some trees were desirable, and prairie settlers and state governments expended money and effort to plant them. Yet the focus on tree planting, which came to be symbolized in Arbor Day, centered mostly around belts and groves of trees and not forests. As Professor J. K. Macomber of Iowa Agricultural College stated in an address advocating shelterbelts: "We don't want forests. The pride of Iowa is in her prairies. If the State were covered with forests, we would cut them down and burn them."[29]

While in Iowa, Bessey engaged occasionally in the debates regarding the scarcity of trees on the western prairies and the climatic effects of trees. H. H. McAfee, professor of horticulture and a colleague of Bessey's in Ames, was a strong advocate of the position that forests increased rainfall and moderated climate, and Bessey was or became a proponent of these ideas.[30] One of Bessey's first impressions of Ames was its barrenness, and he observed that he helped bring "civilization" to the area by planting trees and shrubs on the campus. As to whether trees could be introduced successfully onto the prairies, Bessey started a small plot of trees on the college farm in Ames. In the mid-1870s Bessey was particularly intrigued by the possibility of preglacial conifer forests covering Iowa and by the possibility of planting pines from the Rocky Mountains on the prairies.[31]

It was not until after he had moved to Nebraska in the 1890s, however, that Bessey actively sought answers to the questions of the history and nature of the western prairies and Great Plains. Bessey's establishment of the Botanical Survey of Nebraska afforded a somewhat systematic avenue to pursue these questions. Trees were the topic of two survey reports in the early 1890s and a later one in 1904. The popular part of the first two reports, distributed as bulletins of the Agricultural Experiment Station, provided a guide to tree planting in the state, and the scientific part of each was mainly a catalog that included only the scientific name and a few comments about each species.[32]

As Bessey developed this increasing fascination with trees and the Great Plains, the area that received special attention was the Sandhills, a region covering about one-fourth of the state and commonly perceived as a "barren waste of useless sand" or, at best, of use only for grazing. Of particular interest to Bessey was whether the Sandhills had been forested in the past and could be again. In considering whether the Sandhills had always been treeless as they now were, Bessey stated that "I think it is still the prevailing belief that these hills were always destitute of forests." Most persons believed that trees never ventured beyond the canyons of the larger streams and considered the region a largely desolate, arid desert, a "waste of useless sand thrown up into hills, with no more water than may be found in the hills of the great Sahara of Africa."[33] Despite this dominant viewpoint, there had long been the belief by a few that the region had once been forested.[34]

When first considering the topic, Bessey agreed with the majority that the plains probably had always been treeless but that the prairies may have had trees at one time. By the 1890s, however, he had definitely changed his mind about both areas. His views had changed to the degree that when USDA forester Bernhard E. Fernow lectured at the University of Nebraska in 1891, Bessey persuaded him to provide money and seedlings to test tree planting in the Sandhills if Bessey could find a donation of land. Bessey found a willing donor in his own college when Professor Lawrence Bruner, an entomologist, offered a plot of land in southwestern Holt County situated on the northeastern edge of the Sandhills. The Division of Forestry provided the instructions for the experiment, calling for various plots of different trees planted by a variety of methods, and Professor Bruner's brother did the actual work. The only care called for by the Division of Forestry was to protect the young trees from fire and grazing. Fernow reported the next year that he considered the plantation "established" and that "it seems already to have proved what was intended, namely, that in the sandhill region of Nebraska coniferous growth, especially of pines planted closely, is the proper material and method."[35] Bessey agreed that the test demonstrated that Scotch pines could be successfully grown in the area if seedlings could be kept from dissipating the first year, but it also showed that there was a high rate of loss.[36]

Interestingly, the Holt County experiment appears to have been forgotten for a decade. Bessey later recalled that the "plantation dropped out of public sight and no further reports were made. We supposed, as probably did everybody else who knew of the original planting, that the trees had disappeared and that we had simply one more case of the wreck of tree planting such as were familiar to us in the days of the forest homesteads, known as 'tree claims.' "[37] Such scant attention and significance given to a test in conjunction with the national government seems strange from a man who continued to promote Sandhills forests. Even in 1901 when the reforestation proposal gathered momentum, it was persons in the Division of Forestry and not Bessey who made inquiry into the conditions of the test plots in Holt County.[38]

Bessey continued to inquire into stories of old logs uncovered beneath the sand and in the canyons of the Niobrara River in northwestern and north-central Nebraska, the Loup River in the center, and the Republican River which cuts across the southern part of the state. He wrote to persons in these areas for observations on the distribution of various trees, particularly pines and cedars, and asked how current stands of trees compared with their earlier knowledge of the region or what stories were handed down regarding earlier plant distributions. In 1892 and 1893 Bessey related his ideas on the potential of reforestation in the Sandhills to the State Board of Agriculture and the State Horticultural Society.[39]

A boost for Bessey's ideas on the Sandhills came in an acquaintance with J. C. Toliver, a county judge from Ainsworth in the central part of the state, who had been working with a Timber Culture claim in what he described as some of the "worst type" of the Sandhills. Toliver was enthusiastic about the commercial possibilities for timber in the region and had communicated with Fernow and with the University of Nebraska. Toliver stated that further correspondence with Fernow was useless because the two differed too much concerning the potential of the region. He believed Fernow was mistaken about the water level in the Sandhills, and unlike Toliver—who was certain he could start trees from seed—the chief of forestry thought that only transplanted seedlings had any chance of survival. Convinced that trees could be easily and cheaply grown in the Sandhills by putting seed on uncultivated ground, Toliver felt that fire and not a shortage of water had removed the earlier forests from the Sandhills, a region he claimed was the "natural home of the pine tree." Toliver informed Bessey that he had recently observed an estimated three thousand young pines in a canyon where they had sprouted "just where the seed happened to get a covering" and where they were protected from fire. Likewise, the adult trees he found were in locations safe from fire.

Toliver envisioned that a successful demonstration in the Sandhills would "revolutionize this despised belt" by making it a source of valuable timber, would beautify the region, and would alter the climate by increasing the rainfall and slowing the "strong dry south winds that are such a detriment to our country." Bessey encouraged Toliver to continue his planting tests and provided him with the results of the Holt County experiments and with his own ideas.[40] Apparently Toliver was unable to support his claims. In the summer of 1894 he reported to Bessey that the dry spring had hurt his test of using seed rather than transplants. Still, he wanted Bessey to publicize the efforts and to encourage more people to try similar tests, and in true boomer fashion he assured Bessey that "we feel warranted in saying the plan is a complete success." Bessey seemed not to follow up on Toliver's test, asking in an unrelated letter three years later how the attempts to plant seeds had worked.[41] Nevertheless, stimulated by Toliver, Bessey again raised the subject of trees before the meeting of the State Agricultural Society in 1894 and stated that reforestation "attracted a good deal of attention."[42]

Through 1894 and 1895 Bessey continued to pursue evidence that the Sandhills and other treeless areas of the state had been previously timbered. A curious exchange of letters on the subject of tree fragments buried in the Sandhills occurred between Bessey and Fernow in early 1895. Bessey recalled that Fernow had once told him of finding buried wood in the Loup valley, but Bessey had not written down the account and now needed the information. Fernow denied that he was the source of the account. He claimed that he had obtained the account from Professor Lewis E. Hicks, a geologist and colleague of Bessey's who had helped promote the Holt County experiment. Confused, Bessey replied that Hicks was in Asia but that he (Bessey) was "pretty sure he will tell me that he did not say so as it is likely I would have heard him say something about it. So if you have some other authority, trot him out." If Hicks was not the source, Fernow answered, then he must have obtained the story from Bessey himself. "I don't understand it at all," Bessey replied; "I never knew before that *I* said that *trees were dug up* in Nebraska. It must have been Dr. Hicks. But how did the mistake arise, and how did it escape us until the present?" Bessey further indicated to

## GRASSLANDS AND FORESTS

Fernow at this time that proof of buried trees from recollections of early settlers was not solidly confirmed. Both men undoubtedly had confused Hicks with J. S. Kingsley, a zoologist in Bessey's college. Kingsley had reported in 1891 that he had taken a two-hundred-mile trip across the Sandhills and that preserved pine trunks uncovered in the sand showed that the area had once been more extensively covered than when settlers arrived. As to what had caused the destruction of the trees Kingsley did not offer an explanation, but he judged that climatic changes were not the reason because in the canyons the same kind of pines still grew with no difficulty.[43]

As Bessey continued his inquiry, his correspondence confirmed the stories that the early settlers had dug up pine and cedar logs, at least in the canyons. Bessey apparently dropped the question of buried logs until 1901 when in a letter to Fernow he stated that the forester had raised the matter in 1894–95 and that at the time he (Bessey) had doubted the stories and the "matter entirely passed out of my mind." In 1901 when looking through some of his old letters from around 1890 he found a correspondent who referred to buried logs. This started Bessey again investigating the matter, and he found four persons, including the governor of the state, to confirm that there had been a "good deal of this digging of logs from considerable depth."[44] It is difficult to understand Bessey's forgetfulness regarding Kingsley's account and the numerous references in his correspondence to buried logs and recollections of larger stands of trees at the time of white settlement when one sees the numerous uses he made of the information in the mid-1890s.

The most immediate evidence that the Sandhills had been forested came from tree distribution and migration studies of the Botanical Survey. Bessey concluded that until settlement Nebraska had a much more extensive tree covering that connected the isolated localities where trees now stood. One could not account for the isolated distribution through wind or animal scattering. Although most of the current trees and remains of former trees were of eastern species, some canyons also had western representatives. Bessey found the present and former distribution of the ponderosa or yellow pine particularly revealing. Evidence from stumps and buried logs indicated that the yellow pine had previously extended into the state along some of the major river valleys but had now retreated from all but one. In the interior of the state, widely scattered stands of yellow pine still grew in valleys, and old-timers told Bessey of finding remains of pines between these valleys. "Putting all these facts together," Bessey stated in 1895, "I feel warranted in concluding that the area of yellow pine in this state must have been formerly much greater than it is today. I think I can safely say, also, that it is reasonably probable that the Sand Hills were once wooded with the Yellow Pine."[45]

Further reinforcing his ideas Bessey noted that many pine regions of the world were sandy and "if denuded of trees would be much like our Sand Hills, both as to soil and the surface configuration." He was likely encouraged by reports, such as one in the *American Agriculturist* in 1893, about a region in Denmark that was believed to have been a forest and agricultural area until the twelfth century when it became an "almost unbroken desert." Recently pine and spruce plantations were started on the sand with great success.[46]

Thus believing he had evidence that much of the treeless plains once had been wooded, Bessey then turned to the questions of why most trees had disappeared and whether they were further receding or returning. Earlier Bessey had seemed uncertain as to the historical conditions of the prairies and plains, but by the 1890s he was emphasizing the transitional nature of the Great Plains. Bessey now saw change in the flora of the region in terms of both space and time. The prairie flora attempted to push onto the plains from the southeast while the mountain flora extended into the region from the northwest.[47]

Bessey remained uncertain as to reasons for the retreat and near disappearance of western species such as the yellow pine. When first broaching the subject, Bessey had offered climatic changes as the probable explanation for the retreat. While wind and water currents should favor the migration of western over eastern species, distribution studies showed just the opposite. "I suspect that the meaning of all this," he speculated in 1892, "is that eastern [climatic] conditions are slowly advancing westward."[48] Yet as Bessey continued to gather information during the late 1890s he began to change his opinion about the movement of trees on the plains. He still believed the eastern trees were advancing. (If the young trees were healthy and in advance of more mature specimens, as most eastern species were, he judged that they were advancing.) While his 1895 study of the yellow pine had indicated a retreat westward, by 1900 he had observed advancement by western species in some locations. He also found other species likewise "fighting a winning, and not a losing, battle."[49] In 1900 when Bessey wrote another report on natural migration of trees on the Great Plains, he had made another field trip through the northwestern part of Nebraska and the area north and west of the Black Hills to personally observe the situation. These observations gave him added assurance that in general eastern species, where protected from fire and animals, had advanced over the past twenty years and that western species were now in a position to make similar advances in the near future.[50]

Bessey's certainty about the advancement of eastern species onto the plains was reinforced by other observations. J. W. Blankenship of Montana State College, for example, provided Bessey with information from Missouri, Montana, and Alberta, and in great detail he offered evidence of natural spreading of the entire eastern forest belt in all three areas. Blankenship attributed this expansion mainly to the gradual change in climate and resulting increase in rainfall and only secondarily to the prevention of fires. He acknowledged, however, that his opinions on climate were based more on impressions than on actual weather information.[51]

Bessey earlier had accepted the general opinion that the climate was changing and rainfall was increasing on the prairies and plains, but by 1900 he had collected sufficient rainfall records to assert that "there is not ground at the present time for the belief that the rainfall in Nebraska has increased during the period of its settlement." Likewise there was "no indication whatever from a careful study of the meteorological data that there has been any permanent change in the temperature of the prairie and plains region."[52] Although Bessey still believed that trees and shrubs noticeably curtailed wind along the ground, he now admitted that they probably did not have enough effect on air current above the tree line to alter climate. For Bessey this left the prevention of fires as the primary explanation for the natural spreading of trees that was occurring. Also, fires seemed the best explanation for the original destruction of the forests that had covered the region, with the added destruction of the trees by settlement. Fires not only accounted for the disappearance of forests but explained the remaining isolated stands of trees and their present stands along rivers where canyons would have offered some protection from the burning.[53]

Although he believed that the prevention of fire was the primary factor accounting for the spreading of trees onto the plains, Bessey was interested in the actual mechanism used by trees that enabled them to move. Noting the difference from pines, which to have the best chance for germination scattered seeds onto relatively barren ground, young deciduous trees needed company. The pine spread largely as an individual, but the deciduous plant needed to be part of a community. When deciduous trees were spreading, he observed, "there is always a fringe of weeds and shrubs, and that in this fringe the seedling trees appear." Here he saw a community of trees, shrubs, and weeds, a community in which the other individuals were essential, at least from the standpoint of the success of the trees, but a community in which the individuals were engaged in a "fierce struggle for existence." In further describing this ecological situation, Bessey stated that "here we have a hint that weeds are not always harmful, as in these cases they certainly do serve to cover the ground in such a way that the more permanent vegetation is able to come in after them." Not only did he observe a community at work but also a succession in dominance within the community. "Mingled with the weeds there are always shrubs. At first these are small and inconspicuous, but later they occupy the ground almost to the exclusion of the weeds. In a typical fringe around a forest area the weedy belt is farthest away from the timber, while nearest to the timber is a belt in which the shrubs predominate and where the weeds are of less importance."[54] Bessey thus saw the trees continually advancing upon the grasslands in Nebraska. He believed this advancement was a continuation of a movement that had seen the sixty-five to seventy species of trees found by the Botanical Survey "come into the State, as I think, within comparatively recent times" from the southeast and the northwest.[55]

In 1905 Bessey wrote a summary of his tree migration studies. Following a brief description of wind, water, and soil conditions, he made a short statement about each species. He organized the trees according to the devices (wings, hairs, flesh fruits, rolling balls, edible nuts) by which they dispersed their seeds, and he provided a map for each species illustrating its range within the state. Although Bessey's main attention was on the mechanism by which trees migrate, in a brief introduction he again discussed some of the ecological factors associated with this movement. He noted that plant migration necessitated a "readjustment of the former species, with a necessary change in the relative numbers of the individuals, and the particular habitat of each." Annual plants, he observed, made these readjustments more rapidly than perennials. Yet, he stated, with perennials "there is greater stability, new species finding greater difficulty in entering, and the old species giving away, if at all, only after the lapse of a much longer time." Accordingly, the most "conservative plant communities" were forests and the grasslands of the prairies and Great Plains. Unfortunately for anyone curious to know the extent of Bessey's early ideas on ecology, he did not pursue his discussion of forest and grassland communities further but instead turned in detail to mechanisms of plant movement.[56]

The tree distribution and migration studies made by Bessey and his students were an outgrowth of their scientific curiosity regarding the nature of the prairies and plains, and these studies provided some of the basic information upon which future scientific and applied studies were built. Of immediate concern, these studies provided the basis for Bessey's proposal to reforest the Sandhills, and they provided a significant part of the training for a group of his students who continued to study and to add to the understanding of the forest and the grassland for the next half-century.

With ecology later such an important aspect of the study of forests and grasslands, Bessey's involvement in early American plant ecology is certainly an important consideration of his relationship to these studies. His involvement in plant ecology has long been recognized, but Bessey's interest in ecology was general and usually only incidental to his interest in structural physiology and systematic botany.[57]

In his two articles on seed dispersal Bessey saw ecological questions as a natural outgrowth of Darwinian biology. He considered some community relationships involved in the development of forests. Here he described a community of trees, shrubs, and weeds that formed a fringe in front of the forest, a community in which the individuals struggled for possession of the area. The "fringing belts" provided the cutting edge of the forest as it invaded the grassland. He pictured a progression from prairie grass to weeds, then to shrubs and tree seedlings, and finally to forest conditions.[58] Unfortunately, he considered the topic only in passing. With his belief that a completely scientific survey of a region must bring all aspects of botanical investigation to bear upon understanding the flora, the work of the Botanical Survey had led Bessey into ecological questions of interrelationships among plants and with the physical environment; still these questions did not hold his immediate attention any longer than did other areas of study that attracted his interest from time to time. But now, unlike before, the Nebraskan had a coterie of graduate students who took his lead and moved boldly ahead.

What also seems important about Bessey's attitude at this time is that he did not anticipate or share in much of the enthusiasm generated by ecology at the turn of the century. In his two brief considerations of the dynamics of plant communities mentioned previously, he gave no indication that he felt this type of question was new or of unusual interest. As one of the founders and promoters of the "new botany" movement in the United States, Bessey encouraged an expanded perspective of botany in which all aspects of the living plant—its form, functions, and interrelationships with other plants and with the physical environment—were important. Therefore, ecology, physiology, pathology, and systematics were all important areas of study to Bessey, and he introduced his students to them. Ecology apparently received no special consideration, however, and remained for Bessey a branch of experimental physiology.[59] Bessey strongly opposed too much specialization, even in graduate students, and he insisted that his students have a thorough knowledge of bibliography and research techniques and a familiarity with most subdisciplines within botany. He encouraged his students to study ecology, but he encouraged them to study a broad range of areas and then let their interests and professional opportunities direct them. Such an approach provided good training for ecologists.[60]

Thus ecology was important to Bessey, but during its early development he seemed not to see as much promise in ecology, at least in answering immediate problems, as in some other areas of botany. While he supported ecological studies, much of his apparent coolness toward the discipline at the turn of the century was due to the primitive state of its development, its lack of concise terminology, and the lack of direction among its proponents. Bessey was very critical of many of the increasing number of studies that botanists were producing, some of which he considered faddish, taking common knowledge and presenting it as new under the guise of ecology. As he indicated in 1903, "That is exactly what nine-tenths of the ecological stuff is now-a-days. It is 'haziness and guessing.' Of course, you understand that there is such a thing as good, solid work in Ecology, but there is mighty little of it being done."[61] Another example which demonstrated his dislike for much of the ecology of the day because it was nonexperimental—and therefore of little value—was his suggestion in 1906 regarding the annual field trip of the Botanical Club of the AAAS:

> The fact is that in this day of Ecology, we ought to be able to make the excursion a very valuable thing. If, however, we start out to wade around in some pond or swamp with nothing definite in view, other than picking up a few stray plants and incidentally to wake up the frogs and snakes, little good will come of it. Why not have some of our ecological friends take us out to some place where there is something of real importance? I wish to emphasize the word "real." I do not care to be taken out and have pointed out some very obvious things to which the ecologist has given some newfangled names. I want new facts rather than new names.[62]

Although Bessey regarded himself as an outsider, he was nonetheless influential in training some of the pioneer plant ecologists. Pound and Clements, in particular, built on Bessey's approach to the study of botany, blending as it did the best of laboratory and field, and Bessey certainly was supportive of their approach to ecology. Moving from their work on plant geography (*Phytogeography of Nebraska*), Clements largely provided the quantitative methodology of the Nebraska school of ecology with his *Research Methods in Ecology* in 1905 and the dynamic focus with *Plant Succession* in 1916.[63] Likewise, it was Frederic Clements who introduced ecology as such into the regular part of the botanical curriculum at the University of Nebraska; established the University Alpine Station at Manitou Springs, Colorado; and trained such persons as Raymond J. Pool and Carlos Bates who would continue the Nebraska ecological tradition in both grasslands and forestry. Bessey acknowledged that it was Clements who directed botanists to the "serious nature of proper ecological work," in contrast to most of the "popular" ecology of the day.[64]

During his formative years Clements referred to Bessey as an "ecological sympathizer," and he continued to obtain his former professor's opinions on ecological issues.[65] Although Clements acknowledged that he learned everything he knew in botany from Bessey, in ecology Bessey seemed not to go beyond the general directions he gave in his migration studies, the comprehensive botanical preparation he imposed on his students, and an astute ability to raise pertinent questions and to offer helpful but tentative directions based upon his broad knowledge of botany.[66] Thus Bessey made his impact on students before they took up ecology with Clements or Pool. Also, Bessey's view that ecology rested atop the basics in physiology led him to oppose a move by some botanists to introduce beginning students to botany through ecology. To Bessey this was backward and would only confuse the student as to what botany actually was.[67]

In addition to his interest in trees and grasslands from a botanical standpoint, Bessey participated, again as a comparative outsider, in the youthful conservation and forestry movements, which were in many respects akin to professionalization within the larger botanical community.[68] Bessey shared the view in the 1890s that the nation's resources were fast disappearing with settlement and industrial development. As he wrote in 1894 to his neighbor and congressional representative, William Jennings Bryan, "Our forests are going so rapidly that unless we do something soon there will be none left."[69] Although Bessey had occasionally considered the possibility of state forest reserves and was an early advocate of reserves by the national government, his concern grew in the late 1890s. Attending the Sixth National Irrigation Congress in 1897, Bessey supported the resolution to President McKinley calling for withholding from sale public lands that "are of more value for timber than for agriculture or minerals."[70] Bessey became an active petitioner of Congress when in 1898 opponents of forest reserves tried to reverse the setting aside by President Cleveland of twenty-one million acres of western forest lands.

During the 1880s and 1890s Bessey contributed notes and several articles to forestry journals, and he maintained regular contacts with the Division of Forestry and Bernhard E. Fernow, providing the division with information on Nebraska for its annual reports. He also participated in occasional local and national forestry and irrigation meetings and congresses, but never in a consistent manner. In 1891 he was a member of a committee of the AAAS on "scientific procedures in managing resources." Concentrating on forests and water management, the committee worked for the Yellowstone reserve, but regrettably the Bessey Papers give no indication of his involvement with the committee.[71]

The mid-1890s found Bessey devoting more of his writing to forest journals and meetings. For several years he annually contributed to *Garden and Forest,* and for the American Forestry Association he wrote a summary of conditions of forests and forestry in Nebraska.[72] In 1895 he wrote a feature article in the *American Naturalist* on Charles S. Sargent's study of Japanese forests. From Bessey's tree migration studies he presented a paper to the AAAS, and he summarized his tree distribution work to the British Association for the Advancement of Science when it met in Toronto in 1898.[73]

The contact between Bessey and the USDA Division of Forestry, started under Fernow, continued with Fernow's successor, Gifford Pinchot. In 1899 Pinchot offered Bessey a position of paid collaborator with the division. Bessey was asked to continue his investigations on why trees had disappeared from the prairies. Familiar with some of Bessey's work, Pinchot wrote, "Your name among the collaborators would be a tower of strength and the results which you are so abundantly able to produce would clear up a matter now far from clear in the minds of a majority of men interested in it." Bessey wrote a paper for the division in which he expanded and repeated his ideas about the historical development of the Sandhills.[74]

Bessey served for a number of years as the Nebraska representative (vice president) of the American Forestry Association, and in this position he attempted to organize local support for forest preservation. He was also among a small group that met in his botany classroom in 1899 and organized the State Park and Forest Association with the purpose of "awakening interest in home adornment, city improvement, and planting of parks and forests," and he was elected one of three directors of the new association.[75]

Although Bessey supported forest reserves throughout the timbered regions of the United States, he considered forests equally important to people living in the Great Plains. Stressing the value of forests for water control, Bessey explained that the continued existence of useful rivers and irrigation in his state depended upon protection of their headwaters in the woodlands of the Rocky Mountains.[76]

In the midst of his increased activity in forestry and conservation, Bessey's plans for the Nebraska Sandhills remained his primary interest. During the 1890s he had recognized that reforestation was an immense project and probably premature. He envisioned covering an area 50 to 100 miles wide and 150 to 200 miles long, which would result in a major physical and economic alteration of the entire region. "Perhaps it is hopeless to think of such an undertaking," he pondered; "doubtless it is a project of gigantic proportions, involving, as it does, from five to ten millions of acres." Continuing, he stated that "we are probably yet too busy with the work of bringing the cultivable areas under the plow to consider seriously the work of setting aside lands for forest growth, but I have faith in the ultimate ability of our people to grapple successfully even this great undertaking." Nevertheless he tried to persuade government officials that the area needed to be set aside immediately before it passed into private ownership. He made little headway on the matter, however.[77]

Since establishing the small test plantation in Holt County in 1891, the Division of Forestry's Tree Planting Section had indicated an interest in the Great Plains, but it had not followed up on that test. After the first few years no one even bothered to check the condition of the Holt plots until 1901. At the end of 1900, however, the Division of Forestry picked up the matter of the Sandhills again. William L. Hall, superintendent of tree planting, indicated to Bessey that the division "for some time [has] considered the advisability of recommending setting aside a tract of public land in the Sand Hills of western Nebraska to give the government the opportunity to experiment in a rather extensive way on covering the Sand Hills with forest."[78]

As was not the case earlier, the Division of Forestry was now more serious about investigating timber resources firsthand and considering the possibility of large-scale planting in the Sandhills. In 1901 the Tree Planting Section organized several parties to crisscross Nebraska and survey its timber potential. In the stated purpose of the survey, information for the individual tree planter was still primary, but Hall indicated that his organization might collect information related to forestation and "ultimately perhaps establish extensive government plantations." Because of Bessey's interest and knowledge, Hall immediately consulted the botanist about the planned survey, and while in Nebraska he used Bessey's office as headquarters for the government party.[79]

CHAPTER

Thus in the summer of 1901 many of the leading figures in forestry converged on Nebraska since the survey party was accompanied, for a short time at least, by Pinchot and by F. H. Newell of the U.S. Geological Survey. Bessey moved to take advantage of the presence of government officials and scientists to spotlight his Sandhills project, and he increased his efforts to gain public support. Having established the foundation with frequent writings and speeches during the 1890s, Bessey quickly mobilized the agricultural and scientific societies of the state behind him. The State Horticultural Society, for example, did its part to promote forestry by holding its annual meeting at a time when Pinchot could give an address, and it devoted its entire annual report to forest topics. Moreover members of the society again petitioned the governor and their representatives in Congress to support the Sandhills reserves. Having received the official backing of the state's societies, Bessey approached Governor Ezra P. Savage and persuaded him to assume personal leadership of the movement to obtain the reserves.[80]

What had been clouded with doubt and with little hope of fulfillment during the 1890s now suddenly seemed within grasp. Following the summer survey Hall submitted a report reinforcing Bessey's general claims for the Sandhills. The report asserted that trees could grow and were growing throughout the state. Referring to the Holt County test plot and reports of other lesser tests in the Sandhills, Hall concluded that pines "have grown with great vigor during recent years," thus justifying the proposal that the government should "reserve large areas."[81]

Bessey later recalled that Hall had inquired about the Holt County plot when first arriving in Nebraska. Maybe not coincidentally, it was shortly after Hall's inquiry that Fernow, who was no longer with the Division of Forestry, wrote Bessey to confirm the stories of buried logs in the Sandhills and also asked about the status of the Holt experiment. Bessey assumed the planting to be long dead, and he feared Hall's intended visit there because "its disappearance would be an argument against the possibility of foresting the Sand Hills in spite of any carelessness that might have resulted in the failure of the experiment." Instead of disappearance Hall found tall pines that "had formed a dense thicket in which forest conditions had already appeared." "Pinchot is delighted with what this plantation shows," Bessey related, adding that the trees grew better than if they had been planted around Lincoln. Again Bessey later recalled, "The result of this experiment was to dissipate all doubt as to the possibility of growing pine trees on the Nebraska Sandhills, and, as a consequence, Mr. Hall made the recommendation to Mr. Pinchot that certain tracts of land in the state should be set aside for experimental planting."[82] Interestingly, for the first time in several years Bessey also recalled Judge Toliver's test, but upon inquiry he found that although his trees, which were started from seed rather than transplants, had good germination and early growth they had all died as a result of a long period of drought.[83]

In January 1902 Hall wrote Bessey that "the forest tree planting reserve in the Sand Hills is now almost within grasp." The uniqueness of the Nebraska proposal—setting aside a forest reserve on which there was no forest—posed a possible legal roadblock but one that did not materialize.[84] Implementation required only the confirmation of support from the Nebraska congressional delegation and President Theodore Roosevelt and the final selection of sites. Bessey and the Nebraska Academy of Sciences had suggested several areas. Bessey believed multiple sites were preferable so that they could test a variety of Sandhills types and that there would be less gamble of having to rely completely upon the success or failure of a single, untried location. Hall favored an area of about 1,200,000 acres, but there was little support among Nebraska's members of Congress for such a large reserve. In January 1902 Pinchot sent Bessey a map showing the recommended three sites selected by the Bureau of Forestry. Bessey replied that he was generally acquainted with some of the areas and that he "heartily approved of them," adding that he was impressed that they contained only about one percent private land. He believed there had been enough reliable testing on two of the three proposed tracts that "success of forest plantings is absolutely assured."[85]

With regard to local support, Governor Savage prepared a letter for President Roosevelt with Bessey's aid, and then Bessey responded to the request by Hall for a "strong" letter to Roosevelt with his personal views on the reserves. Pinchot believed the letter from Bessey had "such great force" that it would "have a very decided influence in obtaining the reserves." Stressing that the forest reserves of the Rockies would be insufficient to meet future lumber demands and that few areas suitable for reserves existed in the East, Bessey strongly recommended a policy of establishing large forest reserves on the unsettled Great Plains.[86]

Supporters of the Sandhills had only a short wait until Roosevelt in April 1902 issued the proclamation establishing two reserves, the Niobrara and Dismal River, known collectively as the Nebraska Forest Reserves. As finalized the two sites contained about 85,000 and 123,000 acres. The government did not establish the third area, the North Platte Reserve consisting of 347,000 acres, until 1906 and ended it in 1913. The total of the three was considerably smaller than the originally proposed North Platte tract alone.[87]

In the midst of the enthusiasm over establishment of the Sandhills reserves, Senator Charles H. Dietrich of Nebraska suggested to Chancellor E. Benjamin Andrews of the University of Nebraska that a new department or program in forestry was needed to supplement the new reserves. Sharing the senator's optimism that Nebraska was to become an important timber region, Bessey, as dean of the Industrial College, pledged his support. Likewise captivated by the spirit associated with the reserves, the regents and chancellor immediately established a program in forestry to start in the fall of 1902, and a department of forestry was created at the university the next year.[88]

To initiate the forestry department and to encourage cooperation between the university and the USDA, Bessey proposed a joint appointment in forestry similar to that which land-grant universities had with the Department of War regarding professorships of military science. In support of such an arrangement, Bessey stated that a "man devoting himself entirely to a professorship must necessarily become more separated from technical work and in a few years is out of touch with the work of the Bureau and the work of forestry in general." For Bessey it was also a way to obtain a qualified professor for only half a salary, and the person could spend the remainder of the time for the Bureau of Forestry working at the Nebraska Forest Reserves.[89]

With the bureau's work in tree planting and forest establishment taking place at the Dismal River Reserve, there was potential for the youthful forestry program at the university. As Charles A. Scott, supervisor of the reserves, observed, "I regret to say that I believe forest establishment is not receiving due recognition in Yale. If Nebraska will install a strong silvicultural course I can see nothing to prevent her from ranking second to none for equipping young men for middle West forestry work."[90]

The optimism of 1902 surrounding the founding of the forestry program at the University of Nebraska, however, was insufficient to overcome the lack of financing and other problems that beset the department. From the beginning Bessey encouraged some of his botany students to look to the Bureau of Forestry for careers, as he had done successfully with so many other students in training them for work with the Bureau of Plant Industry in the USDA. But while the "Nebraska element" was becoming "conspicuous" by its increasing numbers in national scientific service, initially those in forestry did not meet with the same success. The training received at Nebraska, which drew praise from physiologists and pathologists with the USDA and which meant that Bessey's students would advance rapidly, was not received with the same enthusiasm in the Bureau of Forestry.

In one of the most interesting cases, Jeremiah Rebmann, a botany student of Bessey's who had developed an interest in forest flora while serving in the Philippines during the Spanish-American War, upon graduation obtained employment with the Bureau of Forestry in early 1902. Placed among forty other student assistants, Rebmann complained that his immigrant, midwestern, and outdoor background ill suited him for office work. In line with Pinchot's warning that persons who were not graduates of forest schools would not pass, Rebmann failed the examination in his attempt to become a full-fledged member of the bureau. Encouraged by his summer work in the field for the bureau, however, Rebmann entered the Yale Forest School in the fall of 1902. At Yale he was no happier than in Washington, D.C. He resented having to work at menial and low-paying jobs to finance his schooling, and he felt out of place around the Yale students whom he regarded as stuffy and aristocratically arrogant. By early 1903 Bessey received a number of bitter letters from his former student about conditions at Yale and the Bureau of Forestry. Declaring he was quitting forestry, Rebmann told Bessey:

> Send the biggest sports you got to Washington and you will have good success. Although Mr. Pinchot is a man of common sense and a gentleman in every way, the rest of the fellows down there (or most of them) are rather sports and like something sporty. They got use for a lot of common sense fellows and hard workers. But they have only a few places that pays a fair salary and the men in knee breeches are bound to get them.

Despite these feelings Rebmann returned to Yale, graduated, and continued to work for the bureau.[91]

Bessey experienced a similar case with another of his students the next year. C. W. Edgerton likewise received his first taste of forestry as a student assistant in the Bureau of Forestry. But upon obtaining a fellowship in botany at Clark University he informed Bessey that "as for entering the Bureau of Forestry again, I would not care to do that.... Under the present conditions in the Bureau, a man has but very little chance to rise unless he is a Yale graduate. I didn't see that the Yale men knew anymore about forestry than many of the student assistants, yet they held all of the higher positions. This is especially true of the Division of Forest Management."[92]

In 1915 the *Forest Club Annual* of the University of Nebraska listed seventy-two persons as alumni of the forestry department. Despite the early foreboding, more than thirty of these were employed in the Forest Service, while several others were teaching forestry or working with state forest projects. Bessey occasionally addressed the Forest Club, and forestry students continued to take botany courses with him and ecology with Frederic Clements or Raymond J. Pool. By 1914–15, however, the forestry department numbered only about forty students, and a problem with the head of the department seemed to coincide with the decision of the regents to drop the program. Its demise came in May 1915, several months after Bessey's death.[93]

While Bessey's plan to convert the Sandhills of Nebraska into a forest and to make the state university a center for the training of foresters eventually fell short of expectations, his hope for trees on the Great Plains was largely successful. Bessey kept close contact with the Sandhills reserves only during the initial stages of tree planting. At the request of the Bureau of Forestry, Bessey reported on the soil, noted some conditions to look for in selecting locations for plantings, and recommended experiments to determine the effects of grazing on the vegetation. While outlining some of the ecological studies necessary for scientific forestry, Bessey did not intend to conduct the experiments or gather the information himself. These were jobs for properly trained foresters and for botanists such as Clements and Pool.[94]

In addition to his purely botanical interests in grasses and trees, Bessey thus played a pioneering role in early efforts to protect and manage the nation's natural resources. He set the stage for much of the work in grasslands ecology and management that later centered in the University of Nebraska. In addition to helping bring trees to the Great Plains, he demonstrated that conservation and forestry were as important to the economic life of that region as they were to the heavily forested areas of the country. Concerned with waste and exploitation, Bessey believed that scientific knowledge was the basis for protecting as well as expanding these resources. While acknowledging that scientists were only on the threshold of fulfilling this role, Bessey pointed out some of the general directions in which botany needed to move to be more valuable. He provided his greatest service by attracting intelligent and energetic students into science and giving them the type of training that enabled them to move into prominence in understanding and maintaining the grasslands and forests of the United States.

# 7

# The Scientist as Progressive: Reforming American Society

CHARLES BESSEY liked the word *progressive*. He used the word frequently, long before it came to designate a political movement or a period of American history. Historians since have seen the Progressive movement from many perspectives and have attached meaning to a very diversified set of ideas and persons. In its most general form progressivism is described as merely the first two decades of the twentieth century. To those historians who see a pattern and unity to the movement, however, it was comprised of the responses, particularly by native-born, middle-class Americans, to increased immigration, to the growth of corporate power and wealth, to the Populist movement (although not necessarily Populist ideas), to the degeneration of the political system, and to the seemingly more numerous technical problems confronting modern society. The picture that emerges is that there were a number of Progressive movements rather than one movement, but central to the persons associated with the movements was their support of reform. Another significant aspect of progressivism was the large-scale involvement of the expanding professions, and Bessey was an active participant in the political and social reevaluations that captivated Americans at the end of the nineteenth century.[1]

To Bessey progressivism was an attitude, a set of values, and an approach to life that he could fit to any set of problems in any time period. Yet like so many others, he displayed a more acute awareness of the political and social issues around him during the twenty years from 1890 to 1910. He divided society into the "progressive" and the "nonprogressive," the "thinking" and the "nonthinking," the "far-seeing" and the "short-sighted," and the "forward-looking" and the "backward-looking."

He believed that societies, like individuals, needed to look ahead, to "think out new things," to favor new ideas.[2] Using a botanical analogy, Bessey's favorite way of expressing this viewpoint was his saying, "Let your brain always contain much meristem [growth tissue], and little permanent tissue and have a bias in favor of new ideas."[3]

Bessey of course placed himself on the progressive side of society. He believed people and societies must continuously grow, change, and improve, and he saw himself as a person who, while not radically discontented with America's past, recognized that change occurs and that people and institutions must adjust. Late in his life he summarized these feelings:

> I do not believe the world half as bad as some people try to make out. There are some bad spots in it here and there, but in general it is a pretty good old world and I have no fault to find with it. When the good Lord made it he seemed to do a pretty good job, and it looks pretty small for us to be making quite so much of a racket in criticism of what he has done. All the same I am a reformer. . . .[4]

Within botany Bessey always saw himself as a promoter of change, and the same was true when he turned to the social and political side of his activities. The changes that were transforming botany into a respectable science encouraged Bessey to join other professionals in pushing science into the mainstream of American life. This push increasingly brought him into contact with politics and politicians and revealed an interesting side of the botanist. Although Bessey never formulated these views, his letters reveal a generally consistent set of social and political values and beliefs.

Throughout his life he regimented himself with the belief that hard work and rigid self-discipline were good for a person. For example, Bessey once apologized to a student that his botanical laboratory would close for two hours on Thanksgiving afternoon because he promised that time to his family. Likewise Bessey believed in strong moral convictions. Religious but not fundamentalist, Bessey had attended the Baptist church in his student days and in later life was a Congregationalist.[5]

Personally he saw little or no latitude for moral laxness, and he had little toleration for such compromises within the society at large, particularly by persons responsible for public institutions such as government or education. He stated, for example, that "the social standard of the college should be set upon a higher plain than that of society at large."[6] He believed the university should carefully control the activities of its students, particularly its female students. When writing recommendations for students he typically indicated whether they were good Christians.[7] In the general community, he publicly advocated the legal prohibition of tobacco smoking, alcohol consumption, and prostitution. In 1898 in response to his articles in the local newspaper against the growing number of taverns, the Lincoln Anti-Saloon League offered him its presidency, but Bessey declined claiming he was too busy. Another growing community concern of Bessey's at this time was the state fair. He complained to the State Board of Agriculture, which was in charge of the fair, that the annual event was becoming less a showplace for agriculture and too much horse racing and games of chance.[8]

Another value Bessey fostered was patriotism. An advocate of the concept of American mission, he enthusiastically supported the Spanish-American War. During the war he occasionally scratched notes or slogans displaying his imperialistic feelings in the margins of his letters. Bessey opposed independence for the former Spanish possessions because he believed the native populations totally incapable of self-government. To Bessey none of the parties involved in the fracas—whether Spaniards, Cubans, Puerto Ricans, Hawaiians, or Filipinos—could lead their countries into the modern world as well as could the United States.[9] Later in 1909, referring to Cuba as a "robber's nest" because of ingratitude for American help, Bessey stated that "I am beginning to be a believer in the ultimate necessity of our taking possession of the Island and holding it as a province until the Cubans learn to govern themselves, which may be in two or three thousand years. Even as to that I am a little in doubt."[10] Aside from the political situation, Bessey looked upon these islands as areas of great potential for botany, agriculture, and forestry. Through two of his former students who were associated with agricultural experiment stations in Cuba and Hawaii he kept close contact on the scientific penetration of these areas.

Over the years Bessey built a political orientation in which he added to the aforementioned basic values certain philosophical concepts that he considered necessary for good government. A believer in hard *money,* or "honest" money as he called it, he attributed to the constant editorials of the Omaha *Herald* the confirmation of his views on the subject.[11] Another facet of his thinking was his opposition to protective tariffs. Bessey combined a philosophical inclination toward free trade, which he developed during his student days, with a practical need to oppose tariff increases. Equipment—particularly German microscopes, books, and journals—was an essential import for American scientific growth. Not until the turn of the century did Bessey judge American equipment an adequate replacement.[12]

Bessey was also a strong proponent of representative government.

Representation combined elite leadership with public responsibility. In viewing society as sharply divided between "thinking" and "nonthinking" elements, there was a strong sense of elitism in Bessey; for example, he saw himself as a trained individual capable of good judgment but believed that not everyone possessed equally good judgment. Yet at the same time Bessey did not think himself significantly different from other largely rural, hardworking people of the American Midwest, good people whose rewards in life were largely nonmaterial. To Bessey these were the kind of people who constituted, or should constitute, the public that government leaders should represent. Thus, he contended, good government was the product of strong but responsive leadership.[13]

In Bessey's view a good government supported the political and moral values he thought essential to any society, but the ultimate test of a society was whether it had a progressive orientation and was governed by progressive leaders. Obviously, then, elected officials needed to come from the progressive, thinking, and farsighted element of society, and to Bessey it was the predominance of such leaders in the Republican Party that gave it an advantage over other political parties.

Bessey had desired to participate in the Civil War but had been too young; he was not yet sixteen years old when the war began in April 1861. Throughout his life he looked upon the wartime Union General Ulysses S. Grant as a model leader because he was a man of action, a man with "grit" who "pushed things." By the turn of the century the brightest star on the political scene in Bessey's opinion was Theodore Roosevelt. Bessey portrayed Roosevelt, who became president in 1901, as a man of action like Grant, and he believed Roosevelt had that farsighted view of America's needs that came only from the strong nationalism and commitment to public service that he was convinced Roosevelt possessed. Bessey was particularly enthusiastic about Roosevelt's effort to curb the power of railroads and to support forestry conservation.[14] In 1901 Bessey lectured his congressional representative: "What the President said in regard to forestry is exactly what I should have said had I been President. We have an admirable leader in the President."[15] On another occasion he claimed that Roosevelt "more that any other single man in the United States is voicing the demands of the people." Bessey concluded this praise by observing, "I do not think that President Roosevelt is so much smarter than the rest of the people in Washington, but he certainly does 'think straight' and if the Congressmen were all to get down to some real good thinking of the Roosevelt kind, . . . the country wouldn't have to wait for something to be done."[16]

Bessey was a dyed-in-the-wool Republican all his adult life, yet he viewed himself as very independent in politics. He always stressed that good representatives of the people should not be too devoted to political parties. When a person gave total devotion to one political party or to any organization, Bessey believed he lost his freedom. "I will act for myself," he emphasized. "I cannot allow others to bind my actions."[17] Periodically he threatened in disgust to leave the Republicans, but he never did. He could always find the alternative less desirable. The Democratic Party, according to Bessey, had too many machine politicians and was too heavily infested with the nonthinking and noncritical minds in the country.[18] In scolding his congressional representative Bessey stated, "It is not half as bad to have Democrats doing nothing because that is what they commonly do, but to have Republicans who are supposed to be in Congress to carry out the wishes of the people doing nothing at all is to my mind quite deplorable."[19] The minor parties, however, received Bessey's greatest scorn. Bessey's most direct dealings with politicians had been with the state legislature over the operations of the university, and it was in contacts with Fusionists, Independents, and Populists that Bessey displayed his bias that leaders of these groups were largely among the nonthinking and were mostly shortsighted.[20]

With society needing what he regarded as the right type of citizens and leaders, Bessey saw education and science as major factors in training progressive minds. According to Bessey, a blending of the traditional cultural courses—such as history, language, and philosophy—with the new sciences would produce more of the desired farsighted citizens. Cultural studies would introduce students to the traditions and values of society, and the sciences would train students to think and to solve problems and should provide the basic principles from which practical solutions could be found to overcome the roadblocks confronting modern society. Bessey believed that modern society faced increasingly technical problems, problems that were rapidly altering America, and the product of this combined education would be the individual whom he considered necessary for the United States to advance.[21]

In addition to proper training, Bessey maintained, progress depended upon members of the thinking element within society contributing their abilities to the nation's welfare as active citizens, government advisors, and civil servants. Bessey had always been a willing correspondent to government on issues in agriculture and science. Although he opposed a national university, Bessey pushed for an enlarged National Herbarium.[22] With the creation of the Carnegie Institution, the question of how the money could best be used to promote science resulted in the usual array of proposals. Bessey was very emphatic that they should not fritter the money in small amounts nor move in all directions.[23]

Increasingly by the turn of the century Bessey broadened the scope of his advice through frequent letters to government officials. Although the Bureau of Plant Industry and forest conservation remained his primary concerns, he pushed for more national effort in promoting scientific research, education, public health, and rural revitalization.[24] Bessey increased his support of the civil service system as a way of bringing well-trained experts and persons devoted to public service into the government. As he noted, "These persons [in the civil service] now work for the public rather than the party."[25] Bessey repeatedly asserted that government scientists were daily demonstrating how a progressive government could serve the public through the application of impartial, scientific research.[26] Not only were these persons becoming experts but they also were solving problems that would direct the nation's future. As he emphasized to Senator J. H. Millard of Nebraska, twenty-five years ago the USDA had been the "laughing stock of scientific men" and now it was the "greatest center of scientific investigation of agricultural problems in the world."[27]

Although a philosophical free-trader, Bessey set aside these laissez-faire attitudes when considering the role of government. While opposed to too much bureaucracy, he generally supported a government with numerous service functions. He believed that state boards of agriculture, state horticultural societies, and state fairs were all legitimate recipients of public money as long as they served the public and not special interests. Despite his continuing efforts to develop science on the state level, however, his experience convinced him that the national government was the key to best serving the interest of the general public. The national government, according to Bessey, was the only level that could have the detachment from local and special interests and the long-range commitment, which he believed states repeatedly demonstrated they lacked, to tackle problems with the magnitude of plant diseases and the conservation of natural resources.

While demonstrating a consistent confidence in what he thought was necessary to advance the country, likewise as Bessey looked at the United States in the early twentieth century there was little doubt in his mind about what were the major threats to American progress. Bessey had enough experience to observe that too often the persons he considered nonthinking nevertheless attained positions of leadership. In addition, Bessey believed that self-serving politicians and special interests were growing stronger. He made the following observation:

> The curse of politics today is that people desire public service for what is in it for them. This attitude of politicians is so general. . . . Unfortunately the virus which has produced the disease of hankering for office for office's sake and not for the sake of the public is so widespread that perhaps what I say sounds visionary. . . . It is because of the abuses which have come into the public service that the party of Lincoln, and Grant, Garfield and Harrison is now so sorely pressed. We must have men in office to do our work and not to further their own personal ends.[28]

In looking at special interests, Bessey shared with many other Americans at this time the belief that political bosses and machines, political parties, and corporations were serious threats to representative government. As a supporter of the Civic Federation of Lincoln, Nebraska, a reform group, Bessey opposed machine politics that he believed had dominated his city. He advocated the direct primary because it removed power from "strict party men," and Bessey considered James Garfield another model politician because he believed Garfield had supported or opposed legislation on its merits and not because it originated with a Democrat or a Republican. In 1909 Bessey supported the insurgent, or progressive, Republicans in their opposition to the Taft Old Guard because he believed they were good representatives who put the interests of the people ahead of private and party concerns. On the opposite side, Joe Cannon of Illinois and Henry Teller of Colorado were prime examples of politicians who Bessey thought placed party or private interest over public interest and were the targets of the insurgents.[29]

In regard to corporations, Bessey favored capitalism and the economic use of the country's natural resources; he was not a wilderness advocate. Yet he believed private gain should come in conjunction with, and not in place of, the general welfare. His personal letters had occasional references to such items as the "tyranny" of railroads and the "dangers" of monopolies. "Railroads are good things," he stated, "but they are good as servants and not as masters."[30]

Bessey's scorn for special interests came forth most dramatically as he continued to advocate the expansion of national conservation efforts. Bessey blamed party-centered politicians and corporations for the depletion of American forests.[31] In 1898 when opposing attempts to restore western forest reserves to the public domain, Bessey stated that this is "solely in the interest of certain corporations which have no care or concern for the future and which are willing to see the water supply of the plains cut off and utterly destroyed as a result of this present personal gains."[32] Because "trusts" and corporations were guided by private and short-term financial gain, Bessey contended, they were therefore improper guardians of the nation's natural resources. The forests, land, water, and minerals of the country were rapidly disappearing under private control, Bessey believed. "We have been unwise, as you well know," he wrote, "in throwing open practically everything to settlement. This country has retained very little for itself."[33] Similarly when he tried to help save some of the giant redwood groves he lamented that "there is no great money interest behind a bill of this kind." Passage of such bills must become a "matter of sentiment or pride in our country and patriotism," but such sentiment, he protested, obviously was lacking in Congress.[34] Regarding the shortsightedness of Congress in not establishing new forest reserves more quickly, Bessey blamed members of congress "who are not especially in favor of anything that does not bring grist of some sort to the corporations."[35] Later in stressing that government forest reserves did not negate using these resources, he stated that a reserve "simply withdraws them [forests] from destruction by private parties who are not interested in the future."[36] On another occasion he added that

> in the management of the natural forest lands of the country we have squandered what should have been a rich heritage to be transmitted to our children and our children's children. The wild forest of the United States originally belonged to the whole country; in other words, they were Government lands and the Government should have adopted such a policy toward these lands as would have conserved them for constant use for all time. The cutting away and total destruction of the forests is a crime against the community as a whole.[37]

Bessey's concern with special interests rose to a pitch of outrage in 1908 and 1909 when again Congress failed to strengthen conservation and forestry policies regarding Appalachian reserves. In 1908 the governor appointed Bessey to the Conservation Commission of Nebraska, and Bessey used this group to continue support for new Appalachian reserves. At the same time they encouraged their representatives in Washington, D.C., to enlarge the bill to protect the eastern slope of the Rockies, which was essential to the protection of Nebraska's waterways. Failure of the bill prompted Bessey's condemnation that Republican leadership deserved abandonment. But again he rationalized that although Republican policy on conservation was misguided, Democrats had no conservation policy whatsoever and "Pops" and Independents lacked the intelligence to carry out a policy if they had one.[38]

Aside from representatives being too devoted to the welfare of their party and to private interests, Bessey believed that the lack of appreciation for science was responsible for much of the seemingly shortsighted action by Congress when it came to issues of conservation and forestry. The application of newly developing scientific principles would, Bessey explained to his congressional representative, make the "forests productive and at the same time keep them alive." He added, "There is no more reason for the destruction of the forests in order to secure the lumber than there is for the destruction of the orchard in order to secure its products."[39] Still fuming over the refusal of Congress to expand forest reserves, Bessey angrily stated, "I wish all members of Congress, especially the Senate, could have injected into their anatomies some knowledge of what real forestry is."[40] Such irresponsibility, in Bessey's view, demonstrated that too many members of Congress were often not only do-nothings but also know-nothings. "It does seem to a man outside," he again observed, "as though we manage to get more dunder heads and slowpokes and men who do not know what to do and do not know how to do it in our Congress than is reputable for us as a people."[41]

Despite his threat to abandon the Republicans in congressional and local races in 1908, Bessey was strongly committed to William Howard Taft for the presidency, though he would split with Taft within the next two years. A friend who was equally upset with the Republicans asked Bessey about William Jennings Bryan as a "citizen and neighbor." The friend inquired, "Is he sincere and honest? . . . I cannot believe that he is simply a cheap demagogue as some report. Shall I vote for him or stay away from the polls?"[42] Bessey replied that he and Bryan were neighbors, belonged to the same club, and frequently exchanged social visits. "So I know him well," replied Bessey, "and like him personally very much. But—in all these years I have learned his weaknesses very thoroughly. And so have his neighbors. We never turn to him to help us solve our public problems. He is not considered by many of us as a man of such sound judgment that we turn to him for advice." Bessey concluded that Bryan was an expert at "making political capital" and that "party expediency dominates with him and has done so from the beginning of his career."[43]

Unfortunately, Bessey's political correspondence greatly lessens after 1909, and he did not discuss his reaction to the formalization of the Progressive Party or to Theodore Roosevelt's Bull Moose challenge in 1912. Bessey never encountered another issue like the forest reserves that stirred him again to political activity.

Thus through this increased interest in government and politics Bessey demonstrated a perspective of American society similar to many other reformers of the late nineteenth and early twentieth centuries. He shared a tradition going back to the "good government" reformers of the 1870s and 1880s, a tradition which stressed honesty, service, and high moral standards in government and which opposed the leaders, but not necessarily the objectives, of groups such as the Populists during the 1890s. He shared the belief that only through reform and change could a nation progress and that in light of America's past the necessary change need not be radical. Personally Bessey was an extremely active person who could not tolerate inactivity in others or by society in general. He could not tolerate waste—whether it was wasted time or wasted natural resources.

His viewpoint, which considered change as necessary, was sharpened by his being a scientist. He saw science as a discipline of action. The history of science was a history of progress, a history of advancement in the understanding of how nature operated, and a history in which it had always been necessary to push aside reactionaries before scientific "truth" could prevail. Proper scientific education, according to Bessey, trained the mind to inquire, to seek answers to problems, and to be impartial in the solution of problems. There was no room in science for the nonthinking or for the preconceived, self-interested judgment. This belief in the expert who could solve the problems of modern society became an important ingredient in the Progressive approach to reform. From the present perspective Bessey's faith in the scientific method may seem simplistic, but the present perspective can never fully capture the feeling of confidence and optimism that resulted from participation in the dynamic upsurge that took place within American science at the turn of the century. Bessey, more than many of his scholarly peers in botany, believed Americans must put this growing scientific knowledge to work for the betterment of society. During Bessey's lifetime he perceived that science had made unprecedented progress, and although generally optimistic that Americans were taking advantage of that progress, he concluded that too often the nonthinking, the self-interested, and the private interest prevented the political side of American society from keeping pace with science.

Although Bessey ventured into the general issues confronting Americans in the new century, science continued to be at the center of his activities. He had reached the stage of his life where he hoped to focus his work more narrowly, and following the wide range of scientific, teaching, and administrative activities during the 1890s and early 1900s,

Bessey was somewhat successful in curtailing his workload after 1905. He rejected offers to assume the chancellorship of the University of Nebraska, having again served as acting chancellor in 1899. He no longer directly handled the duties of botanist for the Agricultural Experiment Station, and he now had more assistance with the teaching in the department of botany.

In 1903 Bessey even took time for a European trip to visit his son Ernst, who was a graduate student at the University of Halle in Germany, and he accompanied Ernst on a research trip through Russia to the Caucasus Mountains. He noted having an interesting afternoon in Moscow with Charles S. Sargent of the Arnold Arboretum and with the conservationist John Muir. Not only did he have an enjoyable trip "botanizing and observing the forests," but his observations were the basis for several papers and articles.[44] He also combined teaching and vacationing during several summers at the University Alpine Station in Colorado Springs, the Minnesota Seaside Station on Vancouver Island, and the Puget Sound program of Washington State College.[45]

Particularly in agriculture Bessey greatly lessened his activities. He addressed such meetings as the new American Breeders' Association and the Dry Farming Congress but diminished his work with the State Horticultural Society and largely stopped his close association with the State Board of Agriculture.[46] His major interest in agriculture apart from the USDA remained education, but even here there was little new.

Bessey continued to be disturbed that agricultural education, specifically as it related to botany and experimentation as conducted by the stations, was still behind the times and insufficient in advancing its potential contributions to agriculture. In 1896 before the botanical section of the American Association of Agricultural Colleges and Experiment Stations (AAACES), Bessey complained that many classes still restricted their study of botany to plant description and identification and were not going beyond Gray's *Lessons* and *Manual*. "Botany of this kind," Bessey exclaimed, "is almost exactly what should not be taught."[47]

Based upon such sentiments, in 1902 the section of horticulture and botany of the AAACES established a committee to study the proper type of botany needed in the agricultural colleges. Headed by Albert F. Woods, a former student of Bessey's who was now with the USDA, the section added Bessey to the committee as an advisor. For the committee and in several later writings on the same topic, Bessey lamented about how little of the "new botany" was used in teaching agriculture. Bessey stated that previously "little more was asked than a classification of the plants used in the field and gardens, with something as to their origin." Bessey added that the older stress on plant identification was "what gave rise to the courses in Economic Botany which flourished a generation ago, and continued Systematic Botany [largely concerned with flowering plants] as the leading feature of the science in the agricultural colleges, and those state universities which gave attention to agricultural education."[48]

This type of education, Bessey believed, did not provide a student with training to do modern work. "In fact," he added, "it was no proper kind of foundation at all, and the men who took up this study were obliged to start at the beginning and develop it for themselves." While Bessey certainly did not want to discard courses in economic and systematic botany, he saw that for agriculture the primary challenge to botany was in pathology. Therefore, to be properly prepared, Bessey stated that "the investigator of plant diseases must know what are the normal functions of plants; in other words, he must be first of all a plant physiologist before he can be a plant pathologist. . . . There can be no scientific treatment of plant diseases unless the effect of the parasitic organism upon the host is understood." In addition to pathology, the rapidly developing fields of scientific plant breeding and agronomy were also dependent upon physiology. Thus, Bessey concluded, physiology had to be at the heart of any botanical research or educational program related to agricultural science.[49]

Following this approach Bessey, as he had done so many times, outlined what he believed was the proper sequence of study in botany. Morphology and taxonomy provided the necessary preparation for physiology, then physiology opened the way to study such applications as pathology, horticulture, breeding, and agronomy. An opposing view to Bessey's, particularly strong among the new professional agricultural scientists, was the belief that agricultural students could learn all the science they needed in their agricultural classes. To counter this Bessey stressed that this approach tended to train narrow specialists. His approach—giving the agricultural student three years of basic science courses followed by one year of applied courses—he believed created more flexibility and enabled the student to apply the principles of science to ever-changing problems and situations.[50]

Another concern in agricultural education at this time was rural schools. Bessey encouraged the normal colleges to prepare more and better teachers and to offer agricultural instruction in the rural and secondary schools.[51] Also regarding rural schools, in 1911 the governor appointed Bessey to the Nebraska Rural Life Commission. In line with President Roosevelt's earlier efforts with the Commission on Country Life in 1908, chaired by the botanist Liberty Hyde Bailey, representatives of the various agricultural societies and organizations asked the Nebraska legislature to follow the example of several other states and appoint a state commission to study the problem of declining farm population. The governor selected nine commissioners, among them Bessey and George Condra from the university, and the commission organized around ten aspects of rural life including health and sanitation, land tenancy, labor, transportation, taxes, and education. Bessey was in charge of the section on agricultural or rural schools.

Regarding rural schools the Nebraska commission was concerned that having to board students in towns during the school year was an important factor in drawing rural children away from the farm. To counter this perceived problem, the national trend was to encourage consolidated schools aided by public transportation. In writing the report for his section, however, Bessey supported keeping the small rural school. Having retained a sentimental attachment to the old one-room school, he even favored such a model for city neighborhood schools in place of the graded system in use. Citing the desirability of individualized instruction and the ungraded, open school, his section proposed a radical concept of rural teaching, yet one that was generally similar to Bailey's idea of fitting the school more completely into the community: they would transform the old schoolhouse, usually isolated and deserted for much of the year, into a community educational center. The school building would become a house for the teacher, a place of instruction throughout the year on an individual basis, a type of model farm for use by the teacher and for instruction, and a community library. The teacher would become a permanent member of the community and be a practicing agriculturist, thus helping to insure a rural education and to overcome the fear of many farmers that the influence of teachers encouraged children to leave rural life.[52]

As in agriculture, Bessey greatly decreased his direct involvement in forestry following the establishment of the Nebraska Forest Reserves and his tree migration studies (see Chapter 6). He did work to extend the idea of forest reserves in Nebraska through efforts to protect the little genuine forest still standing along the Missouri River, and he remained a strong advocate for the expansion of national forest reserves.[53]

Another of the ways Bessey continued to support forestry, and significantly helped to bring ecology and forestry together, was his promotion of his former student Frederic Clements. During the early 1900s Bessey tried unsuccessfully to persuade the Bureau of Forestry to subsidize some of Clements's ecological research on the forests of Colorado. Bessey informed bureau chief Gifford Pinchot that he "should like to see [Clements] turn his talents into such directions as would bring his knowledge of the plant covering of the mountains to the aid of forestry."[54] However, George B. Sudworth, chief of the Division of Forest Investigation, replied that the "so-called Ecology" of botanists might be valuable for forestry, but he believed ecology was "nothing more or less than what we include in our silvicultural studies." Sudworth added that the bureau was already engaged in similar studies and had not yet determined their practicality.[55]

Bessey persisted, however, and in 1907 again stressed to Pinchot the value of Clements's ecological research. Bessey explained that Clements had undertaken his investigation solely to "study the ecology of the forest." Bessey noted, "At first, [Clements] had no idea of its economic outcome, but as the years have gone on, he has found that the work has come nearer and nearer to that which has been found necessary in the Forest Service. Now why cannot these two interests come together?" Pinchot responded by appointing Clements a special agent of the bureau.[56] The appointment coincided with an agreement by Clements to assist the Bureau of Plant Industry in research in forest pathology. In 1908 Clements wrote to Bessey that he was formulating for Pinchot "an ambitious and comprehensive outline for the scientific work of the Forest Service for the next decade or so. This is a very 'large order,' but a very gratifying one and I hope to put the plans in such attractive form that the whole Service will be thoroughly committed to proper ecological methods."[57]

While less directly involved in forestry during this time, Bessey did increase his efforts in the related area of conservation. In 1904 Bessey was surprised by his election to the presidency of the Wild Flower Preservation Society of America, succeeding fellow botanist Elizabeth Britton in that position. He also became involved with the Nebraska State Council in connection with the Outdoor Art League of California, which was conducting a national campaign to remove the Calaveras Big Tree Grove of sequoias from private ownership. Through both groups he hoped to save from destruction a small but disappearing part of America, but legislation to preserve the two groves of redwoods failed at this time and for nearly thirty years.[58]

Another area in which Bessey's work lessened significantly was the flora of the Great Plains. In 1908 Bessey tried to generate student interest and financial support to renew the Botanical Survey of Nebraska and publication of the *Flora*. Following the initial four reports during the 1890s, only three additional ones were printed before the program ended in 1904. The efforts in 1908 were unsuccessful, and it was not until 1917 that the *Flora* was renewed. It is interesting that the first new volume by J. E. Weaver and Albert F. Thiel expanded upon the topic of fringe areas between grasslands and forests, a topic Bessey had discussed in his tree migration studies.[59]

An area in which Bessey remained active during his mature years was educational reform, in terms of both botany and the general nature of public instruction. These activities varied greatly. Earlier he had helped organize educational sessions at the Trans-Mississippi Exposition in Omaha in 1898 and the Congress of Arts and Sciences at the Louisiana Purchase Exposition in St. Louis in 1904.[60] In particular Bessey in his later years continued his efforts to direct the nature of elementary and secondary school botany teaching. With the great changes taking place within American botany, questions of reforming instruction were reaching the secondary and even the elementary levels. How and when botany should be taught were questions for educators as well as for botanists. Most of the debate centered around such issues as separate botany and zoology courses or a combined biology course; field work or laboratory; emphasis on the flowering plants or a survey of all plant groups; collection and identification or physiology; structure and physiology or ecology; and cultural or practical.[61]

Regarding the national concern for public schools, Bessey's involvement had started in the early 1890s when the question of science in several of the nation's large urban school districts became an issue. More generally there appeared to be no order and unity to secondary curricula or to college entrance requirements. Hoping to achieve national reform, the National Council of Education of the National Education Association called for a Committee on Secondary School Studies to organize conferences for each secondary subject. The result was the Committee of Ten, headed by President Charles W. Elliot of Harvard and composed of six college presidents and four secondary school principals. The Committee of Ten established nine disciplinary conferences and appointed ten persons to each one. In the sciences, mathematics had its own conference; physics, chemistry, and astronomy comprised a second; and the third was natural history, defined as "biology, including botany, zoology, and physiology." The result of all this was an 1894 report from the Committee of Ten that was very influential in initially setting national standards for elementary and secondary education.[62]

The natural history conference, which met in Chicago in December 1892, had a strong botany representation of Bessey, John M. Coulter, who was now president of Indiana University, and Douglas H. Campbell of Stanford. The remainder of the ten conference delegates included three other professors—from Williams College, the University of Virginia, and Bridgewater (Mass.) Normal—three high school principals, and the superintendent of schools of Washington, D.C. The natural history conference participants structured their proposed study of biology so that it would be similar in nearly all aspects of rigor, time allotted, and methodology to physics. They recommended a continuous, nontextbook study of botany and zoology throughout the elementary years and at least one full year of botany or zoology at the high school level with the "absolute necessity" that up to three-fifths of the time be devoted to laboratory. The botanists prevented the year of study from being designated more broadly as a biology course, which they saw as essentially zoology. The conference added subreports that outlined high school courses and methods of study for zoology and botany and an elementary course for botany.[63] Coulter wrote the botany part for secondary schools and Bessey for elementary. Bessey was particularly concerned that the teaching in the public schools remained very outdated, and he hoped for changes similar to what he and others had been pushing in the universities. This meant emphasizing laboratory work instead of merely collection and identification as well as the study of all groups of plants in proper lower-to-higher sequence.[64]

Along the same line in 1896 the National Education Association (NEA) organized a roundtable on teaching science in the high school. Charles S. Palmer, president of the NEA, asked Bessey to deliver an opening statement as well as an address to one of the general sessions. Bessey's address, "Science and Culture," described the type of science instruction that was necessary for the secondary student and attempted to show how this instruction had much more to offer the student beyond scientific information.[65]

That same summer efforts by Palmer to involve the AAAS resulted in the forming of a committee consisting of a representative from the American Association for the Advancement of Science (AAAS), the NEA, and each of the five regional educational associations. Instructed to "unify" requirements for secondary science, Bessey represented the AAAS. As expected the committee agreed on the need for science in the secondary schools and generally agreed that these courses should emphasize more laboratory work. The next issue that emerged was whether there should be science instruction on the elementary level. Noting the growing support for "Nature-Study," Bessey stated that "in botany the work of the secondary schools must be based on what has been done in primary and grammar grades," and he concluded that "we must give to nature study much more of value than it has possessed." Bessey outlined a course of study for grades three through eight that was essentially what he had included in his *Elementary Botany* for Nebraska schools several years earlier. As a result of these activities, Bessey later served on the board of the new journal, *Nature-Study Review*.[66]

Locally Bessey established standards for secondary botany. At this time the University of Nebraska exerted tremendous influence over public teaching, being responsible for accrediting the high schools, and during the 1890s individual faculty had to oversee instruction in the schools. In order to be certified in botany, for example, high schools had to follow strict guidelines issued by Bessey, and this included not only books and materials but also such detail as the number of specimens and proper mounting and labeling procedures in a plant collection required of each student. Bessey resisted relaxing the required collection for fear teachers would substitute "book work" instead of more laboratory and field study. Fortunately for the university faculty, after 1897 the university hired a full-time inspector. Even so, as late as 1913 Bessey claimed he still could — and did — control high school botany through the high school inspector who remained a professor at the university.[67] To prepare teachers to teach the type of botany that Bessey deemed acceptable, he taught a methods course on basic botany during some summers. He also actively pushed to reform high schools by requiring teachers to be college graduates, a reform that came to Nebraska in 1907.[68]

Bessey set high standards for secondary botany, as demonstrated when he submitted requirements in Nebraska to a committee of the Society of Plant Morphology and Physiology. The committee, headed by William Ganong of Smith College, was to define what constituted a preparatory course for college botany. The committee found Bessey's course too advanced for the average secondary teacher as well as the student. Another issue surfaced when the committee disagreed with the Nebraskan's organization. Bessey's secondary course mirrored his college course, and starting with the "lower plants" meant working at once with the microscope and with cells. The committee believed students could better handle general plant structure before tackling cells. Another major disagreement between Bessey and the committee was the growing issue of ecology.[69]

This was not Bessey's first encounter with the place of ecology in the beginning botany course. Following their work with the NEA during the early 1890s Bessey and Coulter began to part ways on the proper methods of teaching secondary botany. Again as members of an NEA committee in 1899 Coulter outlined a sequence of teaching that placed major emphasis on ecology as featured in his new textbook, *Plant Relations*. Bessey greatly opposed this. Ecology, Bessey believed, was not proper for introducing students to botany. The base for understanding plant science had to be structure and a general introduction to plant groups, then physiology. Only at this point should students be introduced to plant relations and communities. Emphasizing ecology also meant stressing field work rather than the laboratory. Noting that he was "dreadfully tired of the fad," Bessey felt that Coulter's use of ecology was putting the cart before the horse and would confuse beginning students as to the nature of plants. He believed ecology was for advanced students. The other committee members, however, agreed with Coulter. Labelling the committee's work as "quite absurd," Bessey refused to support the report.[70]

Like the NEA committee, Ganong's committee also strongly disagreed with Bessey regarding ecology. "The ecology germ," Ganong observed, "is in the system thoroughly of the teaching botanical public, and probably no course in the near future could be acceptable which did not include a considerable portion of it." Judging Bessey "unfair" in his condemnation of how ecology generally was used in the classroom, Ganong concluded that "it seems to me a very attractive kind of knowledge for students."[71]

Attempting to reflect the changes in American botany in teaching at all levels, efforts by these committees demonstrated the difficulty of agreeing upon what was needed other than change from the old emphasis upon collection and identification of the flowering plants. These disagreements continued in subsequent years but with less passion.[72] In 1910 at a symposium regarding the need to train more and better botany teachers—participated in by O. W. Caldwell of the University of Chicago; Frederic Clements, now of the University of Minnesota; F. C. Newcombe of the University of Michigan; Bessey; and Coulter, now of the University of Chicago—there was no more agreement than before on the state of teaching and on what was needed for its improvement.[73]

But they kept trying. Bessey correctly characterized this as a "period of educational unrest," and he, as did others, continued to expound upon what was proper teaching of science and upon how the schools could become more progressive.[74] But even in his presidential address to the AAAS in 1912, Bessey lamented that despite all the growth of his discipline over the past four decades botany "is yet largely unorganized and lacks consistency in plan and purpose," adding that "there is yet no agreement even upon so small a question as to the content of the first year of college botany, or the mode of its presentation." On this point he concluded that "it is to our shame as botanists that we acknowledge our inability hitherto to frame a standard first-year course in college botany. When the science is definitely formulated in the minds of botanists the present disagreement will no longer exist."[75]

On this issue, as others, Bessey had difficulty conceiving that thinking people would reach different solutions. When "experts" disagreed Bessey generally believed there was a right and wrong, so someone's thinking remained faulty. Coulter, on the other hand, was much more tolerant that there was not one proper way of what and how to teach. Good teachers, Coulter believed, could develop a variety of approaches and each could be equally successful. Interestingly the two had few differences regarding the organization of college botany, and even on the secondary level they had more in common than disagreement.[76]

During the first fifteen years of the twentieth century as Bessey continued efforts to mold the institutions in which he had had so much influence over the past three decades, he looked pleasingly on the professional changes within botany and science in general. Likewise he was proud of his part in securing a greater role for the USDA and agricultural science generally. But closest to home, at the University of Nebraska, some of his main ideas related to the organization of the Industrial College and particularly the place and role of applied education began to come unraveled.

Within his own department of botany there were few problems. Enrollments remained good and the department was strong in graduate as well as undergraduate instruction. Bessey retained his control over the botanical work in the agricultural experiment station and the agricultural preparatory school even though he became less and less directly involved. He remained the ex officio botanist of the experiment station, but the actual work was done under his direction by other botanists and assistants from his department. In 1904 Bessey reorganized the department of botany, placing the work in agricultural botany and the research of the experiment station completely in the hands of one of his assistant professors who also handled the advanced teaching of pathology and some of the physiology. In addition he turned some teaching of physiology within the department to Frederic Clements, who already was responsible for ecology. These changes allowed Bessey to focus on systematic botany in addition to his administrative duties.[77]

Unrelated to these organizational changes, but of interest, was Bessey's selection of Elda Rema Walker as a replacement for Clements when he left the University of Nebraska in 1907. A new Nebraska Ph.D., Walker had been an instructor before her promotion to adjunct professor. Later joined on the botany faculty by her sister, Walker spent her entire career at Lincoln.

While women had been a normal part of the land-grant college, only recently had they begun to move outside the separate programs established for their sex or outside the arts and sciences college (at Nebraska the sciences had been in the Industrial College). Yet before 1900 there were few women in advanced botany at the University of Nebraska. Bessey always had women correspondents—mostly former students, teachers, and amateur plant collectors—but none were in his inner circle of academics as they planned and promoted the "new botany." Also, at least in his correspondence, Bessey did not challenge the prevailing attitude that elementary or secondary teaching, or such an exceptional position as a scientific artist, were about all that an advanced female botanist could expect in terms of employment.

Bessey always referred to "his boys," for example, when considering career possibilities in the USDA. Carrie Harrison, a Nebraska graduate in the 1890s, learned firsthand that stenographer positions were all that were offered to women when she applied at the USDA. In 1904 when Harrison joined four other of Bessey's students—Albert Woods, Herbert Webber, C. L. Shear, and Ernst Bessey, all of the USDA—in presenting a framed photograph of Bessey to display in the office of the Division of Vegetable Pathology and Physiology, she was then working at the U.S. National Museum. Adeline Ames was the first "Nebraska girl" to obtain a scientific position with the USDA, becoming a pathology assistant to Erwin F. Smith in 1905. Ethel Field, another pathologist, followed in 1909. By the end of Bessey's career in 1915 when seven of the twenty-one forestry pathologists in the Bureau of Plant Industry were Nebraska graduates, Ruth Bates Fleming was one of the seven.

So a change in botany was occurring by 1905, and although Bessey was a contributor to that change and certainly favored it he did not publicly acknowledge the growing presence of professional women botanists until his last major address in 1914. Even in 1912 in his presidential address to the AAAS all his discussion regarding the future of botany had been in reference to what men must do. Yet two years later at the twenty-fifth anniversary of the Shaw Gardens in St. Louis, Bessey recognized that "this banquet was notable in that for the first time there were women among the guests, as should be, of course, when we remember the very considerable number of women who are engaged in botanical investigation, and in botanical teaching."[78]

Another interesting development on the university scene in the early 1900s was the emerging issue of college athletics. The sign over Bessey's desk, "Football occupies the same relation to education that a bullfight does to farming," openly displayed his general dislike of the growing intercollegiate competition. Favoring physical education for the general student body, Bessey complained that what had been entertainment and a weekend diversion was spreading into the school week and conflicting with classes and particularly afternoon laboratories in the sciences. He was certainly wrong when observing that the sports craze would quickly end and when he stated that "this is merely a phase in the development of college management and eventually we shall bring colleges back to their legitimate work again. We are now in the stage of athletic domination." While unsuccessfully haranguing the chancellor about abolishing intercollegiate sports, or at least deemphasizing them, Bessey joined the growing ranks of persons whose concern even reached the pages of *Science,* concern that resulted in national action to reform football and the intercollegiate scene in general.[79]

Of more immediate concern to university organization, in 1906 in opposition to Bessey's continued efforts to keep the agricultural programs fully incorporated into the regular university structure and to share faculty with the academic departments, the university separated the agricultural experiment station. The botanist of the experiment station and the department of agricultural botany was no longer part of Bessey's botany department. This was in line with what had already been done in chemistry.[80]

Also at this time efforts increased to change the Industrial College. Pressure for change came from several directions and had been mounting since the late 1890s. In part the pressure was due to the success of Bessey's college. Following the reforms of 1889, when all the academic science classes were moved to the Industrial College, enrollments had increased. In addition there was a noticeable move of male students from the arts college to the Industrial College, particularly into the emerging engineering programs.[81] Pressure also came from another direction. Engineering, followed by the general science course, came to dominate the Industrial College, but the agricultural programs within the college continued to languish. Charges from outside the university again surfaced, claiming the university was still treating agriculture as a second-class program and demanding either total separation from the university, a separate campus, or an enhanced and separate program within the university, that is, a College of Agriculture.[82]

By the turn of the century, professionalization in both engineering and agriculture led the faculty of these disciplines within the Industrial College to desire their own administrative operation and their own courses. Under Bessey the programs in engineering and agriculture remained quite structured, basically three years of general science courses. The engineers and agriculturists wanted less general science and more application, plus they desired to teach more of their own foundation courses. Finally rumblings of discontent mounted as faculty of the arts college saw enrollments slipping and more of their potential students going into the Industrial College. In 1906 for the first time enrollments in the Industrial College surpassed those in the arts and appeared to be continuing steadily in that direction. The arts faculty faced this unsettling change at the same time that the regents stressed placing more priority on the applied programs. The arts faculty protested and the administration agreed that the arts and sciences college should be the primary academic focus of the university. These various pressures resulted in a major restructuring of the colleges in 1909.[83]

In 1908 in his report to the regents, Bessey had recommended the changes that were in line with the views of Chancellor Samuel Avery and the regents. The Industrial College was abolished. The programs in agriculture and engineering each went their own way into new separate colleges, and the basic sciences went back to the College of Literature, Science and the Arts. Bessey was now a dean without a college. As the administration searched for an appropriate title for the botanist's new position of vice chancellor, senior dean, or head dean, Bessey commented that he would now have more time for his botanical work. He also was pleased that the courses of the College of Agriculture would be maintained on an equal level with the other university applied programs, and the college would not be separated from the university.

Bessey now stressed that the proper structure of the university was three-tiered, with the arts college providing the general foundation, the applied colleges (teachers, agriculture, engineering, law, medicine) constituting a second level, and the graduate college being what he called the third or university level.[84] Probably only Bessey's status and insistence had kept the old structure intact for so long—twenty years. Certainly Bessey could see that specialization would have the same splitting effect within the university that it was having in the scientific societies. While professionalization was bound to affect the perspectives of the new people coming into the applied areas, Bessey saw more harm than good in the changes within agricultural science. The growing emphasis on application and the lessening on the basic sciences was contrary to Bessey's lifelong approach of how to train applied scientists.

# 8

# Evolution and Classification: The Bessey System

AS VICE PRESIDENT of the American Association for the Advancement of Science (AAAS) and chair of the Botany Section, Bessey in 1908 delivered an address on "The Phyletic Idea in Taxonomy" in which he outlined his current ideas regarding a natural classification system. Generally this was a progress report on ideas first stressed fifteen years earlier when he had delivered a similar vice presidential address to the very first meeting of the Botany Section. Although these ideas had been discussed since 1880, Bessey's 1893 address had been his first major statement about his plans to bring plant classification in line with evolution. Despite his many activities and varied interests, the one topic to which Bessey repeatedly returned during the 1890s and beyond was classification.

While acknowledging the central role of evolution in botany, Bessey paid little attention, at least publicly or in his correspondence, to the controversy over the specific mechanism of evolution. In his few references to the workings of evolution, Bessey consistently adhered to the idea of natural selection, particularly to the concept of the struggle for existence. Like most Darwinists he accepted that variation occurred naturally in plants without being greatly concerned about the mechanism that originated the changes. This mechanism was a key problem to neo-Lamarckians, who believed that changes in organisms occurred as an adaptation to the environment and were passed on to successive generations under the French naturalist Jean Baptiste de Lamarck's theory of the inheritance of acquired characteristics. It was variation that was the basis of natural selection, a process in which those organisms best suited to their environments were the most hardy. Bessey saw nature working in this manner as trees tried to compete against grasses on the prairies and plains. In another case he saw in the historical emergence of the angiosperms the crowding out of the earlier types of plants. Bessey contended that the angiosperms came to dominate the world because of their "extraordinary plasticity" and because they had "developed every possible device to survive the fierce competition." Even in his last textbook in 1914, in regard to variation Bessey only discussed natural selection and Mendelism, mentioned mutation theory, and ignored acquired characteristics. Reviewing the book, Byron Halsted criticized Bessey: "I regard that in a textbook where evolution is the key, the subject of species-making is not presented somewhat more fully and historically." But he added that "it is judged that the authors [Ernst Bessey joined his father as coauthor] are essentially Darwinians who strengthen their book by frankly stating their ignorance of the way 'inherited variations' arise."[1] Despite his preference for natural selection Bessey, while frequently discussing evolution, seldom referred to the controversy over mechanism and the increased criticism of natural selection.

What was true of Bessey was also true of many botanists. Although most made occasional references to the issues of neo-Darwinism, neo-Lamarckism, and other proposed means of descent, no heated public discussion occurred the way it did over nomenclature or even teaching. There was very little direct discussion of evolutionary mechanism in the *Botanical Gazette,* for example, particularly before 1905. While there was considerably more on the topic in *Science,* little of it was by botanists. The same was true for the *American Naturalist,* which remained under the control of Edward Drinker Cope and Alpheus Packard for over two decades starting in the late 1860s. These leading American neo-Lamarckians certainly dominated the evolutionary discussions of the journal, and although Bessey was in charge of the botany material for most of the time Cope and Packard edited the *American Naturalist,* he completely ignored any discussion of mechanism. What little botanical material appeared on evolution was general and did not discuss the pros or cons of natural selection or acquired characteristics. Likewise in their correspondence Bessey and Cope never discussed what could or could not appear in the journal, and there was no indication by Bessey that the difference in their views on evolution was a problem.

But even in the *Naturalist* and particularly in *Science* the multiplicity of views and the growing confusion over the issues of how evolution worked were evident. Of the two main groupings in America, the labels of Darwinism and the increasingly popular neo-Lamarckism covered a growing range of issues and viewpoints. A variety of other viewpoints, including neo-Darwinism, recapitulation, geographic isolation, organic selection, and orthogenesis, further complicated the scene by the 1890s.[2] Examples of this range of discussion by botanists include the following: In 1893 Frank Lester Ward delivered one of the few specific addresses on "Neo-Darwinism and Neo-Lamarckism." John Macfarlane, who was interested in cell structure, reviewed the evidence supporting acquired characteristics and—contrary to general opinion—stressed that plants provided more evidence for this viewpoint than did animals. W. J. Spillman and H. J. Webber, both of the USDA, discussed "Environment As a Factor in Natural Selection" before the Botanical Society of Washington in 1902, with Spillman stressing the role of environment and Webber taking the opposing side. O. F. Cook opposed both natural selection and acquired characteristics, instead favoring his kinetic theory, or concept of vital motion, as the force behind evolution. F. E. Lloyd and LeRoy Abrams responded on opposing sides to zoologist David Starr Jordan's proposition that isolation was the primary factor in creating new species. Finally, reviews of the numerous books on evolution contained occasional comments regarding factors for or against one of the contending schools.[3] Generally in these discussions natural selection remained the point of departure. In the Spillman-Webber debate, for example, Spillman ventured that natural selection was "overworked" since most variations were neither beneficial nor harmful and hence of no influence. Webber deemphasized the importance of environment, believing that environmental-induced variations were not inheritable and seeing the influence of the environment as destroying through natural selection those plants that did not adapt.[4]

Apparently for most botanists the topic of evolution was becoming too large to understand more than a small part, if even that. Even for those few persons who engaged in debate on mechanism, often referred to as "philosophical" evolution, increasingly the argument was not that only one method was correct but that one method was more correct or played a greater role than another. It was also evident that the relationships among the different groups, the individual authorities, and the issues associated with each were not clear-cut or unified.

An interesting summary of American evolutionary thinking appeared in *Johnson's Universal Cyclopedia,* a publication that emphasized science and had a strong group of professional scientists as writers, including paleontologists John W. Dawson and Edward Drinker Cope, zoologists Theodore Gill and David Starr Jordan, and horticulturist Liberty Hyde Bailey. Bessey wrote the botany sections of the new editions of *Johnson's* in 1893 and 1897, but the zoologist J. S. Kingsley of Tufts University did the articles on evolution and heredity. Kingsley's general views on evolution, however, were very similar to Bessey's. Kingsley was still inclined to accept natural selection, stressing the struggle for existence and, unlike the neo-Lamarckians, placing less importance on what caused variation.[5] Kingsley ventured beyond this basic viewpoint and stated that the German zoologist August Weismann's work on heredity was probably the most important since Darwin's *Origin of Species* and provided the major basis for the neo-Darwinian school. In the 1880s Weismann had proposed a concept of heredity based upon characteristics contained in the sex cells and not influenced by changes in the body cells. If true, his ideas would be a major obstacle to key premises of the neo-Lamarckians. Kingsley continued that as a result of Weismann's hypothesis "the philosophical naturalists are arranged in two hostile camps, the neo-Darwinians, and the neo-Lamarckians." While admitting the issue was still unsettled, he believed "the burden of proof seems to be in favor of the former side [neo-Darwinians], and it is incumbent upon the neo-Lamarckians to show some way in which acquired characters may be transmitted, or some flaw in the logic of Weismann." Yet Kingsley recognized the confusion, stressing that the neo-Lamarckians could present "many facts, chiefly from the fields of botany and paleontology, which are seemingly easiest explained upon the hypothesis of the transmissibility of acquired characters." Kingsley, as did most writers, referred to all the basic ideas of the various schools as hypotheses. Most importantly, Kingsley concluded, "it must not be understood that either side denies the truth of evolution. . . . The sole dispute is as to what kinds of variation can be inherited."[6]

Liberty Hyde Bailey was one of the few botanists who directly discussed the murky issues of evolutionary mechanism, and he summarized the confusion nicely. Starting from the premise that organisms are too complicated to conform to any one law, he believed it was unfortunate that most persons tried to account for all of evolution with "some single hypothesis." Rightly or wrongly, Bailey believed that since the premises of Lamarckism came from animals and Darwinism from plants most botanists started as supporters of natural selection and that all recent evolutionary ideas were primarily modifications of, or responses to, natural selection. Thus he was a Darwinist because of what he studied and the way he viewed the problems he confronted, yet he did not believe that persons who studied other parts of the world's flora and fauna were wrong in what they saw. There was good evidence that both natural selection and acquired characteristics were at work in nature. Bailey concluded that "species do originate by means of natural selection, but that not all species so originate." And he added that "I am a Darwinian, but I hope that I am willing to believe what is true, whether it is Darwinian or anti-Darwinian."[7] However, though he accepted Darwinism and neo-Lamarckism as true, he observed that "Weismannism, or the Neo-Darwinian philosophy, may be true for some organisms, but it is wholly untenable for plants."[8]

In addition to such biological problems as variation, adaptation, and heredity, evolutionary concepts were further complicated by their association with philosophical and religious issues. For many observers support or opposition to the various evolutionary schools depended upon the real or implied positions of the group in regard to purpose in nature and design, ideas associated with natural theology and German idealism. In the United States, the attempts by Asa Gray to defend a theistic evolution stand out as an example of this influence, and a major element of neo-Lamarckism had a strong emphasis along these lines resulting in movements entitled organic selection and orthogenesis.[9]

If Bessey is typical, there was no consistency or clarity on where botanists stood on these philosophical and religious aspects of evolution. On the one hand, Bessey stood strongly for keeping religion out of scientific issues. He believed clergy, in particular, by their training along "lines of faith and reverence for authority," were unsuited for science, which he maintained demanded "if not skeptical spirit one of complete independence. Science has no infallible gospel to settle disputes except nature which interpretation is difficult as only the original investigator knows."[10] Thus he believed one should not look to religion for answers or guides to scientific problems. Likewise he was very critical of writers who by the 1890s still held to such ideas as seeing "nature as a perfect system and plan," plant parts as representing a perfect order, or such notions as the nature of thistles and briars resulting from the sins of Adam and Eve.

Yet on the other side, Bessey had read and apparently approved Asa Gray's attempts to reconcile evolution and religion in *Darwiniana*. Also by at least the 1890s Bessey had gained a following among the local clergy in Lincoln, Nebraska, for his lecture to church groups on geology and Genesis, but unfortunately he did not leave a copy of this. Likewise he was able to answer the North London Christian Evidence League in 1909 that he was a Christian who believed "Jesus the son of God," that the Bible "contains a divine revelation," that nature strengthened the concept of an afterlife, and that the "universe had an intelligent first cause." Bessey never seemed bothered by the two sides of such issues, believing that evolution should not affect one's religious convictions and that religion should not play a part in proving or disproving evolution.[11]

By the turn of the century several new areas of research — centering around the Dutch botanist Hugo de Vries' new concept of mutations and the revival of the 1860s work on cross-breeding pea plants by the Austrian monk Gregor Mendel—promised new insights, yet they added to the quandary over evolutionary mechanism. Among botanists in general, mutations attracted more attention from the "philosophical" evolutionists (at least judging from publications in the journals), but among particular groups such as plant breeders, both Mendel and de Vries created great excitement.[12] In 1903 Liberty Hyde Bailey declared de Vries' ideas as the newest and "most pronounced counter-hypothesis" to natural selection, de Vries denying that selection could produce new species through the minute variations of Darwin. Like Bessey, at this point Bailey did not foresee particular significance in mutation theory, regarding it in many respects as merely a "rephrasing of the old idea of sports" or monstrosities. Yet Bailey did believe that de Vries provided a major contribution by setting the example of "making actual experiments the test of [evolutionary] doctrine."[13] Interestingly, early enthusiasts for Mendelism included R. A. Emerson, horticulturist at the Agricultural Experiment Station of Nebraska, and Herbert J. Webber of the USDA—both former students of Bessey. While both were actively involved in plant breeding research, Bessey was slow to warm to genetics. In 1904 for example, he related to Emerson that "as you know I am somewhat of a doubter in regard to Mendel's Law. I think that the fellows have been more excited over it than they need be."[14]

As experimental botany became more of a reality in the early twentieth century, some of its advocates became very critical of the lack of evidence backing many of the so-called "philosophical" evolutionary schools. Daniel T. MacDougal of the Desert Research Laboratory, as a champion of experimental botany in the United States, early accepted the orientation of de Vries and the proof of mutations. MacDougal tended to deemphasize both natural selection and use and disuse, believing internal or physiological factors far more important in species creation than any external factors. By 1908 he maintained that there was only experimental proof of species-creation from mutations, hybridization, and natural selection. Increasingly he opposed the neo-Lamarckians, claiming that their ideas and evidence were "so vague, so inclusive," and eventually by 1911 he called for their proponents to show experimental validation or drop out of the contest. While he was ready to discard the neo-Lamarckians, still he did not believe that experimentation would eliminate the explanations for evolution down to one final cause.[15]

In many respects the experimentalists represented the emergence of another level of professionalization, an attempt to raise the method and subject of botany to a higher plane and to separate themselves from the "new botany" of Bessey's generation. As in earlier cases this conscious effort to separate from the previous level probably exaggerated the differences between the experimentalists and the type of botany they criticized or tried to improve upon, although certainly there were some differences. Bessey sensed that some of the pronounced differences of the experimentalists were merely rhetoric and were exaggerated, but undoubtedly he felt no uneasiness with the laboratory even though his current system-building in plant classification was the type of botany under criticism, a criticism he addressed later.

It was in this context of "philosophical" confusion that Bessey, in the introduction to his 1908 address to the AAAS, placed the controversy involving evolutionary mechanism into the perspective of his work on plant classification. "To-day every botanist is an evolutionist," he began. "It may well be that we have not yet agreed as to the details — as to the particular manner in which modifications were effected — whether they were by slow and almost imperceptible deviations from the parental type, or those more marked variations that we are in the habit today of calling 'mutants.' " Others, he continued, stress more "survival of the fittest," "survival of the unlike," "struggle for existence," "adaptation," "inherent tendency," or "environment." "Yet," he concluded, "with all this diversity of opinion as to details there is a practical unanimity as to the acceptance of the general doctrine of evolution."[16]

For Bessey, as for many botanists, whether truth rested with the Darwinians, the neo-Lamarckians, or the neo-Darwinians was a secondary issue. For the particular work they were doing the acceptance of evolution was enough, the specifics of how variation occurred and how or whether variations could be inherited mattered little. The general acceptance of evolution determined how a botanist viewed the plant kingdom, lower to higher, and viewed structure in terms of modification and descent. "With such an agreement among botanists as to the validity of the doctrine of evolution," Bessey reasoned, "it needs no argument to sustain the thesis that a natural classification must be an expression of a theory of evolution."[17]

. Starting with his early textbooks, Bessey had outlined the necessities of a natural classification system and believed that evolution now provided a scientific basis for making such a vision reality. Over the years Bessey compiled a set of basic principles to govern his view of how plants had evolved. By the last statement of these principles in 1914 Bessey had seven general laws or "dicta." These included such ideas as: although evolution was not always in one direction, once a direction was "set in, it is persisted in to the end of the phylum;" green plants always precede colorless ones; and changes in the various organs of a plant do not occur at the same pace or even in the same direction. In addition Bessey had seven rules relating to the general structure of flowering plants and fourteen relating to flower structure.[18] Ultimately natural classification had to reflect the existence of a genetic relationship between different groups of plants and not merely show structural similarities. Relationships became the key concept.[19]

A natural system had been the intended purpose of most classification since Linnaeus, and leadership in this endeavor came from such key figures as Antoine-Laurent Jussieu, Michel Adanson, August de Candolle, John Lindley, George Bentham, and Joseph Hooker. The ideal was to group together plants that were most alike, and the key to likeness was primarily form. These efforts resulted in major improvements, with de Candolle's work dominating Europe and Bentham and Hooker's influencing classification in Great Britain and the United States. By the mid-nineteenth century, however, two significant developments—evolutionary theory and the rush of German morphology and physiology—placed classification in a new perspective. The major systems in use essentially were all nonevolutionary, and the growing body of botanical research made the basis of relationships of the earlier systems appear primitive. Yet by Bessey's day the old systems of classification still retained much of their authority or at least usefulness.[20]

American advocates of the "new botany" had stressed the need for classification reform since the late 1870s and 1880s. Frank Lester Ward, for example, in 1878 expressed similar ideas to Bessey's on the need for classification to reflect descent.[21] Bessey recalled having first considered the issue when reading the German zoologist Ernst Haeckel.[22] Closely related to the new botanists' concern that classification must recognize the new views was their realization that most systematists or taxonomists, despite their acknowledgment of evolution, continued to treat species as fixed entities, and the issue of the nature of species remained muddled.[23]

In several letters in 1892 Bessey indicated he had ideas to help remedy the situation in classification. Commenting upon a proposed revision in classification by Stanford botanist Douglas Campbell, Bessey said to "wait for my system" and indicated it would be in print later in the year. In a similar communication with Columbia botanist Nathaniel Britton, Bessey noted that he was not enthused about the recent work of the German botanist August Eichler and asked Britton to "wait until what I have worked out comes to you. I should then like to have a review of 'Bessey's system.' "[24] The promised publication was an outline of his proposed revision of plant groups appearing in his *Botanical Exercises,* but his first major statement on classification reform was his address to the Botany Section of the AAAS in 1893. As time for the meeting approached Bessey wrote his former student Byron Halsted that "I hope to make a radical address. Get ready to back me up. I may need it."[25]

In his address Bessey chastised fellow botanists for their slowness. Starting from the premise that "all botanical knowledge finally culminates in some kind of classification," he emphasized that the "greatest grouping of all" is not physiological or morphological but is a complete picture of the plants of the world, laid out in full schematic fashion, showing their relationships and descent back through time. Bessey reminded his listeners that it had been a third of a century since "a great light was first turned upon all biological problems" by Darwin. "I need not now," he stated, "before a body of scientific men, speak of evolution as an hypothesis: for we know it as a great biological fact, as to the existence of which there is no shadow of doubt." Bessey emphasized that "we now know what relationship means." Evolutionary concepts provided "new light," new pathways, and gave new meaning to the ideal of a natural system. In evolution, he continued, "we have a principle of classification worthy of modern science, but a practice which abandons or ignores it." Bessey judged that the major classification systems following Linnaeus, from Jussieu and de Candolle to Bentham and Hooker, in most respects remained very artificial. Certainly since the conversion of botany to evolution, he believed, more progress should have occurred, yet most Americans still had not progressed beyond Bentham and Hooker's views of the plant world of the 1850s–1870s.[26]

Why had botanists been so reluctant to devise new classification systems in light of evolution? Bessey believed that the reluctance was primarily an indication of the entrenched conservatism that dominated the discipline. This conservatism was reflected in the hesitation to move without the guidance or approval of the dominant persons who had regarded the floras and manuals as their domain, that the Hookers and Grays of the world were the only ones who possessed the authority to act on such grandiose schemes. Also there was a conservatism that said that not enough was known, that botany must wait for every group to be studied in the detail necessary to reveal natural relationships, and that all the details must be in place before botanists could venture any generalizations. Part of this conservatism was a result of hesitancy on the part of botanists to speculate. Botanists had been working to be accepted as scientists and some of the experimenters, Bessey observed, had confused speculation and nonscientific generalities with the bold general view that every science needed to advance. This reluctance to change when faced with the need to do so was to Bessey the sign of a conservative mind and not the mind upon which to build a progressive science. The challenge

Bessey presented to his listeners in 1893, and was to repeat many times thereafter, was to abandon these old systems and to begin immediately to replace them with new systems—systems more reflecting evolutionary relationships.[27]

By the time Bessey issued his challenge to Americans, reforms of the old systems of de Candolle and Bentham and Hooker were already appearing, particularly from Germany. In the 1870s and 1880s the German botanist August Eichler outlined a system that, while not evolutionary in intent, generally fit into the new perspective and began to replace de Candolle in Europe. Many of Eichler's ideas were quickly adopted into American practice. Starting in 1887, however, the major German thrust came when Eichler's work was reinforced and modified by Adolf Engler and Karl Prantl with the appearance of their summary of plant families in twenty volumes and eleven editions and of Engler's *Syllabus of Plant Families* beginning in 1895. Additional influence on American classification came at this time from a variety of botanists, including Hans Hallier and Karl Goebel of Germany and Eugen Warming of Denmark.[28]

In addition to Bessey, others in North America began to ponder reform of the major plant groups during the 1890s. D. P. Penhallow of McGill University and Douglas Campbell of Stanford, for example, used Goebel as their model. The most influential American convert to German reforms was Nathaniel Britton, who announced in the middle 1890s that the proposed multivolumed and multiauthored *Flora of North America* would follow the general outline of Engler and Prantl. Coupled with the adoption of Engler for Britton and Brown's *An Illustrated Flora of the Northern United States, Canada, and the British Possessions* and for the AAAS Checklist of Plants in 1895 these changes, to Frederick Coville and the editors of the *Botanical Gazette,* "marked an epoch" in American systematic botany. Conway MacMillan of the University of Minnesota agreed that it was testimony to the "passing of the old regime in arrangement."[29]

Bessey certainly recognized the significance of these changes, and while he welcomed reform he did not share the excitement over the use of Engler and Prantl. He believed that changes in classification systems were going to undergo tremendous revision in the near future and that something as significant as the *Flora of North America,* which would probably take fifteen years to complete, would be outdated when finished. Unable to persuade the authors of the flora not to follow Engler and Prantl, Bessey offered as a practical compromise the printing of plant families in some loose-leaf arrangement. He stated to Britton that the "arrangement of the angiosperm families from Engler and Prantl which are an improvement over that of Bentham and Hooker is yet so unnatural in some parts that we ought to be able to rebind the volumes when we get ready to do so." Bessey hoped to delay the rush to the German reformers until he could "have my revised Benthamian sequences so worked that it could be followed by those who see nature as I do."[30] As Bessey explained:

> The study has been, however, a very interesting one to me for many years, and the more I look into the matter the more I am convinced that the present arrangements are decidedly faulty. The arrangement which we have in Bentham and Hooker we all know to be seriously at fault, and that given us by Engler and Prantl is also very illogical. I cannot hope to give to the world a faultless system, but it seems to me that the main features which I have proposed must be in accordance with Nature.[31]

As American taxonomists adopted these new systems, resistance to change came mainly from two directions. As with the controversy over nomenclature also raging at this time, one cluster of holdouts to the move to break the English lock on American classification contained the adherents of the Gray tradition. Bessey, for example, criticized M. L. Fernald of Harvard for not following any of the "more natural systems," and it was not until the 1908 revision of Gray's *Manual* by Benjamin Robinson and Fernald that they brought this American standard into line with Engler and Prantl. Bessey applauded this, stating that the change was in the "spirit of Gray" who had always been "progressive."[32] In addition to the removal of the authority of Gray and thus indirectly Bentham and Hooker, the classification "explosion" of the mid-1890s and after was stimulated on both sides of the Atlantic by a "gush of textbooks" which resulted, according to the editors of the *Botanical Gazette*, in "schemes of classification appearing with bewildering frequency."[33]

In response to this flurry of activity, a second group resisted reform on practical grounds. The problems with a natural system were many and consensus was less with the new systems than with the old. Growing criticism of the Bentham system had caused some movement toward alternative systems, particularly Engler's, and now no sooner were these changes occurring than Bessey and others were asking for an abandonment of these systems. After 1896, for example, teachers using Bessey's textbook faced the problem of their students seeing one system presented in the textbook and at least one or two other systems presented in the manuals and monographs that they used. For many botanists the way to avoid such confusion in teaching, as well as in research, was to use the old system of Bentham regardless of its weaknesses.

The growing attention to classification reform resulted in more discussion of what a natural scheme should be, what would be necessary to achieve such a scheme, and the major problems that would be encountered. Bessey proposed starting such reform by reversing the arrangement of plant groups that placed the flowering plants first in any sequence because they were the most familiar and most important, an arrangement used in America most notably by Asa Gray. To Bessey an evolutionary system should start with the most primitive or earliest organisms and work up to the most mature or most recent. He believed this was true not only in terms of general arrangement—that is, from lower to higher plants—but also in terms of structure and function; investigation of plant structures must start with the primitive and be able to demonstrate the development to the mature.

In practice most systematic botanists concentrated primarily on structure in tracing the modification of parts from group to group. This emphasis upon comparative structure or morphology, while the mainstay of the old classification systems, took on new meaning from the mid-nineteenth century onward, led by the work of Wilhelm Hofmeister, Hugo von Mohl, and Carl von Nageli in Germany. Their pioneering investigations stimulated another generation in Germany centering around Julius von Sachs, Anton de Bary, Karl Goebel, and Eduard Strasburger. The result was a growing understanding of cell structure, vascular bundles, and protoplasm. Reproduction took on new meaning with greater knowledge of pollen tubes, pollen, embryo-sacs, and the concept of alternation of generations. The structure of the plant thus came to be known in ever more detail, and it was this growing detail that the systematist believed could provide a more accurate basis for tracing modification of plant organs from one group to another.

Generally, systematists believed structural differences between individual organisms were most revealing, although they stressed that ideally the investigator should study the entire life of the individual from embryo to adult, in all its internal as well as general structure, and in its functions, habits, and relationship to its surroundings.[34] Despite their protestations to look at the totality of the plant and its environment, most systematists, including Bessey, remained in the tradition of de Candolle and believed reproductive structure held the most promise to reveal relationships. But they could not ignore an ever-increasing abundance of other promising information that provided enough evidence to thoroughly muddy the view and to prevent any consensus as to what in the plant most revealed ancestry. Bessey, for example, stated that the history of the individual and the species "appears much alike" and that recapitulation theory, wherein supposedly the individual embryo retraced the history of the species, was "probable" in providing definite clues to ancestry. By the late nineties he was less cautious and declared it an "established principle" that embryonic stages "resemble" the adult stages of earlier evolutionary groups. At the same time he placed little importance on sports or monstrosities and believed them misleading in tracing descent, an attitude that later affected his reaction to mutation concepts.[35]

Certainly considerable complications marred a successful pursuit of a natural system based upon structure. Particularly confusing were such questions as which flower parts were simple and which were complex: a single or a compound pistil? separate or fused stamens? separate or united flower parts? Bessey acknowledged that the answers to these questions, even after long and careful study, depended very much upon individual judgment.

Just as confusing was the recognition already evident to observers that while nature had generally progressed from simple to complex, it did not always do so. The job of the systematist would have been considerably easier if nature had progressed in a straight line from one group to the next. Unfortunately, as Bessey noted, there were "multitudes of cases" where plants apparently had structurally degenerated, a situation not recognized by Engler. The difficulty in determining genetic relationships, therefore, was complicated by the problem of determining not so much what was simple in structure, but what was simple as a result of being primitive in time rather than simple as a result of degeneration. An evolutionary arrangement needed to identify what was primitive in order to know what came first in time and sequence. It then needed to be able to distinguish the primitive type of simple structure from the simple that resulted from modification and degeneration of what once had been more complex parts. This latter simple would have to be much later in time and certainly was not primitive but advanced as a characteristic.[36]

Already the increasing structural knowledge had produced a quagmire as far as natural relationships were concerned, and it was no wonder that many systematists found it easier to avoid the confusion by continuing to use the old and admittedly defective systems. The allure of a natural system of plant classification was now over a century old, but what looked like forward steps to one generation of botanists looked like a continuation of artificial ways to the next. Still, while a natural system had proven elusive in the past, to Bessey botany now had the conceptual framework and methods at least to begin to remove the obstacles to a more natural classification, and Bessey spent much of the last two decades of his life trying to provide such a foundation.[37]

Starting with three very active years, 1893 through 1895, Bessey decided to lead classification reform by example. He had followed standard classification in his textbooks until 1892 when in his *Botanical Exercises* he ventured to introduce an outline of his own system. Acknowledging his use of Karl Goebel's descriptions of the various plant families, Bessey created his own arrangement and explained that his was better than Goebel's system because it "attempts a natural arrangement" while Goebel "dodges the question."[38] In 1893 Bessey added another outline of his ideas, "Synopsis of the Vegetable Kingdom," which he claimed were "different from anything else in print" and particularly a better arrangement of the monocots.[39]

As early as 1893–94 he devised a chart illustrating the relationships of the major plant groups, intended to help students visualize the relationships of the groups. Criticizing the normal listing of plant groups, Bessey correctly asserted that most illustrations were linear, placing one group after another, and thus did not show the points of deviation of one group from another or properly display the evolutionary patterns. Charts showing the starting points and branching of the various plant groups became a trademark of Bessey's publications and helped gain acceptance of his classification system because of its usefulness in teaching. Much later the branched chart, using straight lines, was replaced with jointed segments. Bessey determined the size of each segment by the number of families in the group, and the shape depended upon the amount of deviation and direction of development. He joined the segments together at points where he believed the departure into a new group had occurred. The result, what came to be termed "Bessey's cactus" because of the jointed nature of that plant, provided a good visual presentation of the main lines of development.[40]

More of Bessey's "new" ideas soon appeared in the introduction to the *Flora of Nebraska* and in *Johnson's Universal Cyclopedia*. Boldly in 1896 Bessey put his own system into his *Essentials of Botany*. Also during the 1890s and early 1900s he presented three papers to the AAAS and a paper and presidential address to the Botanical Society of America on classification of the flowering plants.[41] From 1894 to 1905 Bessey prepared a series of nine studies on the "lower plants," emphasizing their supposed relationships.[42]

The "lower plants" presented many problems, and much confusion remained regarding their relationships. With fewer detailed studies available in comparison to the flowering plants, many life cycles had not been worked out.[43] Therefore there remained much disagreement as to whether various structures constituted separate organisms or merely different life phases of one form. Bessey observed that "lower plants" by their nature were much more "plastic" than the more stable higher forms. Being much more responsive to change and able to change more readily, Bessey added, "lower plants" therefore were much more illustrative of evolution.[44] One of what were to become Bessey's basic phylogenetic principles was that "structures with many similar parts (homogeneous) are lower, those with fewer and dissimilar parts (heterogeneous) are higher."[45] If this were true, determining distinctions of the general plant groups would be most difficult among the lowest organisms. Rusts were a particularly good example used by Bessey to illustrate the difficulty encountered. He pointed out that structurally rusts, like other fungi, were extremely degenerated. Similarity of structure among the various rusts also presented problems, as did the great variety in life cycles. Another reason for much of this difficulty was that it was hard to see these structures with any clarity or consistency.[46]

Because there was little agreement on classification of the "lower plants," Bessey had essentially used his own arrangement since his first textbook in 1880. Few of the old classification systems covered these organisms, and there was nothing for the lower groups in the United States, for example, that was equivalent to Gray's *Manual* of higher forms to use as a descriptive guide upon which to base a classification scheme. In general Bessey worked through each group, usually in light of the special and detailed studies by other scholars. Although descriptive work of algae and fungi increased in the 1890s and Engler and Prantl covered these groups, conclusions remained much more tentative than among their higher relatives. Using these new and more detailed revisions, Bessey attempted to determine what was primitive and then how modification had occurred and where, thus obtaining the points of departure from one group to another.

The points Bessey emphasized in these studies were that the elusive bacteria did not comprise a single group of organisms but had developed from a variety of previously unknown forms.[47] Bessey acknowledged that the origins of these most primitive groups were blurred, as was the distinction between plants and animals. Opposing the English botanist A. G. Tansley's viewpoint that had plants developing from the flagellates, or the German Engler's that had flagellates as plants, Bessey reasoned that plants came first, with animals, including flagellates, being the derivative.[48] He sided with most other botanists at this time in putting the slime molds on the animal side, but recognized that like bacteria most of these simple life forms remained in a gray zone between the two kingdoms.[49] Bessey departed from most systematists, particularly Engler, by not placing the algae and fungi in two distinct groups. He scattered the fungi among the orders of algae, viewing the fungi merely as degenerated algae, but he admitted that botanists really were as yet unable to trace the lines of origin of fungi.[50]

Finally he considerably rearranged the orders of algae even though four major studies had only recently appeared during 1904–5.[51] While basing his evolutionary relationships on the growing detailed studies, primarily of plant structure, Bessey believed that too many of these studies resulted merely in "heated debates over little details of structure" and were of little use. Rather, the purpose of these studies should be to enlighten botanists of the relationships between these structural changes from one group to another.[52]

Despite all these difficulties and disagreements, Bessey moved ahead to outline what he believed to be the major evolutionary relationships and the structural changes that demonstrated those relationships. In 1902 Bessey used the opportunity of his presidential address before the American Microscopical Society to summarize his current views on the "lower plants." Starting with the simple, and primitive, single-celled plants, Bessey moved from the loosely connected filaments, through the undifferentiated cellular plants, to the differentiated cellular organisms. This change in plant life reached to the lower green slimes. As differentiation increased through the pond scums, the major change was reproductive, moving from fission to simplified sexual cells and conjugation. The changes in reproductive structure were accompanied by gradual development of protection for the reproductive cells and intercellular partitions. At this point, Bessey believed the seaweeds diverted from the algae along a dead-end path, and the molds departed in another direction due to their growing dependency (parasitic and saprophytic) and hence degeneracy. From the highest green algae, greater size, protection, and differentiation in the sex cells were more fully apparent in the mosses. In a "short step" from the liverworts to the ferns and then to the flowering plants, he saw the modification of sexual organs and an accompanying general structural complexity.[53] Bessey summarized by observing that "the vegetable kingdom is a unit as to origin, and its multitudes of forms are connected by an unbroken series of evolutions of structure into structure. To the discerning mind there are no exceptions, no forms which are not related to others earlier than they."[54] Following this series of articles from 1896 to 1905 Bessey produced only one other detailed analysis of a lower group, that was a revision of the conjugate algae in 1914.[55]

Although busy with working through the "lower plants" in preparation for his still-promised general textbook in taxonomy, Bessey did not ignore the higher groups. However, he never spent much time on the middle groups of plants — liverworts, mosses, and ferns; generally he accepted these groups as fairly clearly distinguished and not subject to much controversy.[56] The primary interest of Bessey and most others studying the higher orders was the origin of the angiosperms, or flowering plants, and it was the answer to this question that was the key feature of the "Bessey system."

As with the "lower plants," there was no agreement upon or way to verify what was primitive and what was derivative. Bessey's basic ideas about the origin, and hence relationship, of the flowering plants reversed the generally accepted arrangement. The common view had the Ranales (buttercups, peonies, magnolias, tulip trees) as further along the evolutionary path, having developed by reduction of parts from the more primitive but structurally more complex composites.[57] Following his "laws of evolution," Bessey reasoned that the greatest similarity between major groups, such as the monocots and dicots which nearly all now agreed were closely related, would occur in their most primitive stage, before they had deviated very far from each other. Taking both the entrenched Bentham and the increasingly popular Engler systems, Bessey believed the lowest families in each of their groups of monocots and dicots were not much alike, so obviously they were not the original point of divergence in the evolutionary past. By this test Bentham was far from a natural system. Even the Engler system, which Bessey had favored as better by at least placing the simple plants first in the arrangement, Bessey now judged to be "little better as an expression of genetic relationship than the system of Bentham and Hooker, which it is now displacing."[58] Bessey believed that if one took the characteristics of the Ranales in dicots and the families with equivalent characteristics in the monocots and compared them, they would be very similar and meet the test of his law. Thus, to Bessey the Ranales were the key in determining what was primitive.[59]

Accepting the Ranales as his base group, the divergence of monocots and dicots would be those groups most like the Ranales, the Liliales (monocot) and the Rosales (dicot). From each of these diverging lines, Bessey could then rely upon the accumulated knowledge that had gone into the Bentham and Engler systems to see which groups in the monocots and dicots were most unlike the Liliales and Rosales. Here you would find the groups (Asterales of dicots and Orchidales of monocots) having departed the most through both complexity and reduction of structure, the least primitive groups. Using this reasoning Bessey now had the characteristics of the Ranales as primitive traits (separate and simple pistils, stamens, and leaves) and had the advanced traits of the groups at the other end of the process. Over the years Bessey added what he considered primitive nonreproductive traits. Detailed studies of the remainder of the groups between his two opposing ends would fill in the natural paths followed by plant development and result in a natural — or *more* natural — system.⁶⁰

Bessey's confidence in the significance of Ranales grew as he believed his reasoning was backed by his own years of study, but also by the ever-growing detailed work in morphology, histology, and embryology. Bessey thought the findings of these observational and experimental studies generally reinforced what, when first proposed in the 1890s, had been largely a deductive system. In the early 1900s as he significantly extended and clarified his classification system, considerable help came from the growing literature on gymnosperms.

As early as the 1870s, George Engelmann and Lester Frank Ward had related to Americans the inadequacy of the old views regarding the position of gymnosperms. In the older schemes gymnosperms had been paired with dicotyledonous flowering plants, and this pair was grouped apart from monocots. Reformers believed that gymnosperms more correctly should be separated and monocots and dicots grouped, but only gradually did this view come to dominate American thinking.⁶¹ Following this major rearrangement considerable interest remained regarding the exact relationship of gymnosperms to angiosperms and the origin of both groups. Additional work tended further to separate the gymnosperms from the angiosperms and to locate the origin of the gymnosperms in the ancestral ferns.⁶²

New studies, particularly in paleobotany, stimulated additional interest and provided potentially significant new information. Generally botanists had been slow to incorporate fossil work. The fossil record was acknowledged to be an essential ingredient in any evolutionary botany, but most botanists found little clear and substantial information in paleontology. A common complaint was that too much of paleobotany was poor botany. Yet by the late 1890s *Botanical Gazette* editor John M. Coulter, among others, began to stress the need to incorporate more fossil information. In 1897 David White, a paleontologist, pointed out to Bessey the great changes in European paleontology over the past fifteen years, and while admitting that the botanical side of the field was still weak, he believed it could prove valuable at least for the higher plants. Particularly White stressed that there was much of value for understanding the gymnosperms.⁶³ Bessey admitted that even though he had lectured on fossils in his systematic classes since 1884, he needed to give more attention to the subject. But there was little visible evidence of paleobotany in Bessey's writings until after 1900.⁶⁴

Increasingly, however, the direction of paleobotanical interpretations proved particularly helpful in supporting Bessey's classification system. An "epoch making" study, according to Bessey, was Yale paleobotanist G. R. Wieland's 1906 "American Fossil Cycads." Bessey accepted Wieland's views that placed the fossil cycads, or cycadioids, as the point of origin within the gymnosperms for the emergence of the angiosperms. The fossil cycads not only provided a possible solution for the point of origin of flowering plants but also, significantly for Bessey's purposes, this group fit with the Ranales better than it did with the primitive groups in alternative classification systems.[65] Another study influenced by Wieland was by Britons F. A. Newell Arber and John Parkin, who likewise identified the fossil cycads (Bennetitales) as the direct ancestors of flowering plants. Again the structure of the Bennetitales seemingly fit with Bessey's ideas regarding their descendants, the primitive Ranales. The conclusions of Wieland, Arber, and Parkin hurt Bessey's opponents who insisted that the flowering plants had to originate with simple, petal-less types. As Bessey happily boasted, "as a consequence they arrive at the conclusion that primitive angiosperms were necessarily polypetalous, hypogynous and apocarpous, precisely the conclusion reached by me on theoretical grounds more than fifteen years ago, and since then persistently held in the face of the increasing popularity of Engler's system."[66]

Bessey now believed there was research support for the basic arrangement of the flowering plants as he presented them in his "Synopsis of Plant Phyla" in 1907 and his speech to the Botany Section of the AAAS in 1908.[67] In his opinion it was no longer merely a deductive or speculative system. In his address he outlined the history of the development of his system and noted that the intense attention by a growing body of researchers over the past fifteen years had confirmed the basic ideas of his system. The key was the acceptance of his three primitive groups — buttercups (Ranales), water plantains (Alismales), and roses (Rosales). Accepting the characteristics of these groups as primitive, there was general agreement of the groups that must stand at the opposite end of each path diverging from these three groups (two dicot paths, one monocot path). "In fact," Bessey concluded, "when we agree to the hypothesis that polypetalous, hypogynous, apocarpous flowers are primitive the great outlines of the phylum (or phyla) are quite obvious, and the only questionable points are with reference to the place and sequence of intermediate orders." Having presented the "great outlines," Bessey asked his audience to use his system as a "general working hypothesis," a system that he believed would serve "to more clearly apprehend the mode of evolution in the vegetable kingdom, and the consequent relationship of the resulting multiplicity of types."[68] Of course, not everyone agreed with Bessey's primitive groups, and Engler's system continued to have followers in the United States and to dominate Europe.[69]

Despite what Bessey and his followers believed to be convincing evidence, the speculative or deductive nature of his system remained a major point for criticism regarding the validity of his conclusions. If one accepted Bessey's premises, then likely they would accept the major scheme. There would always remain disagreements over many of the more difficult or little-studied groups, and while each revision of his "Synopsis" was an attempt to answer these minor criticisms and to incorporate new studies, the essential part of his system remained intact.[70]

Although providing three more revisions of his "Synopsis" in 1909, 1910, and 1914, Bessey never finished his long-promised general textbook on systematic botany. Unable to complete the textbook, Bessey used the 1914 edition of his *Essentials of Botany* to present his latest thinking on the origin and arrangement of flowering plants. Likewise Bessey's last public defense of his system was in 1914 at the twenty-fifth anniversary of the Missouri Botanical Garden.[71]

The main emphasis of his 1914 address was methodology, and he answered the criticism of the experimental botanists for the type of activity he had pursued for the past twenty-five years. Defending taxonomic system-building as a legitimate scientific activity, Bessey likened this work to detecting ether or determining the nature of electricity, light, or gravity. There was always a need in any science, he asserted, for deduction and speculation. The building of hypotheses based upon what observation and experimentation were available was necessary because educated guesses were needed to guide observation and experimentation. Such work would always be tentative and constantly in need of revision in light of new inductive results, Bessey conceded, but that was to be expected. Waiting for all the facts, on the other hand, would never work to advance knowledge, he claimed, because as in the case of the origins of flowering plants the fossil record probably would never be complete enough to provide "direct evidence." Thus the alternative to system-building, with all its weaknesses, was to abandon the goal of a natural system and to adopt a workable, acknowledged artificial catalog scheme. Bessey totally rejected this alternative for a number of reasons, the most important of which was that to abandon the search for a natural system was to give up the best demonstration available of evolution in action, and evolution, despite all the confusion and disagreement over the past twenty years, remained a pillar of modern botany.

Bessey's classification system never fully supplanted Engler's; still it provided an alternative view, logically conceived and vigorously supported, particularly concerning the flowering plants. Its use in textbooks and by many of Bessey's students, who were well trained in the basic assumptions of the system, assured that it would continue to be used.

But its use went beyond his students and continued to gain support during the 1920s and 1930s, particularly in the United States.[72] A popular account in 1932 proclaimed that Bessey had won the "battle of the buttercups" from the Germans and tried to explain to the public the significance of this controversy.[73]

By the 1950s and 1960s standard American taxonomy textbooks listed Bessey's as one of five to eight general classification systems (along with those of Bentham and Hooker, Engler, Hutchinson, Tippo, Melchior, Cronquist, and Takhtajan) still in use throughout the botanical community.[74] Certainly by the 1950s all these systems were becoming dated. Attempts to base classification on evolution had centered primarily around the same principles of comparative structure in which rank was assigned to various characteristics of the plant. Just as Bessey and his contemporaries condemned earlier systems as inadequate and artificial, in the end Bessey and Engler suffered the same fate, and their period of classification largely ended as different assumptions gained control of plant taxonomy. This fate, however, did not diminish the contributions of that period to our understanding of evolutionary botany today, and rightfully later pacesetters such as Arthur Cronquist, a leading American taxonomist, acknowledged Bessey's role in the classification of the flowering plants. Cronquist observed that "in 1915 Bessey published his epochal paper . . . in which he set forth the principles on which a phylogenetic system should be founded, a list of putatively primitive characters, and an outline of a system incorporating these ideas." Recognizing that "although the system itself is now generally conceded to be faulty in execution, most of his principles (stated as dicta) are widely accepted," and as a result Cronquist concluded, "we are all — or nearly all — Besseyans."[75]

As one of the "spiritual fathers" of the "new botany" movement in America, honors were many during the last stage of Bessey's career. He continued to hold offices in numerous professional and university organizations, and as his generation of scholars moved off center stage he was recognized for his lifelong contributions. In 1906 a group headed by former student Albert Woods pushed for his election as Secretary of the Smithsonian Institution, but the selection did not come.[76] However, Bessey around this time served as president of the Botanical Society of America and the American Microscopy Society and was a vice president of the AAAS for the second time. Finally in 1911 the most prestigious honor of all came to Bessey with the presidency of the AAAS. He was only the third botanist to be so recognized by that body, and fittingly so for Bessey because it was the organization he prized most highly.

Bessey had helped see American botany through a significant period of its development. In botany Bessey's reputation was never based on a dazzling discovery or restricted to a small isolated part of the plant kingdom. Instead it was his ability to analyze and to reject or absorb each new piece of research that made him a botanical clearinghouse of information. Bessey's work seldom described his own experiments or discoveries but included analyses of recent research by others with suggestions of what botanists had overlooked, or not stressed enough, and what research was needed next.

His brilliance lay in his ability to see, understand, and explain the wide range of botany in all aspects, basic and applied. It was the breadth of his view of the plant world that made him one of the leading proponents of the "new botany." He often explained that his academic isolation in Iowa and Nebraska, as well as the seemingly endless demands on a land-grant professor, accounted for his lack of specialization. Yet in most respects his lack of specialization was his strength. During the 1880s and 1890s he had helped guide the direction American botany needed to take. While outlining the need for new botanical avenues and providing leadership for how they should be pursued in both field and laboratory, he left the actual work largely to others. After 1900 Bessey was little attracted to become involved in what the next generation believed were the areas that would provide a "new" botany emphasizing genetics, ecology, and experimental physiology. Bessey certainly kept abreast of the changing scene in twentieth-century botany, but outside classification he was not an active contributor. His students were busy, however, and while Bessey's classification system for flowering plants would be his most visible legacy, the numerous students throughout the broad range of botany, agricultural science, forestry, and education proved to be his most lasting contribution to American science. He had done his job; American botany had significantly matured, and now new ventures would build upon that level of growth.

Bessey died on February 25, 1915, and a scholarship fund in his honor was quickly initiated by alumni.[77] But the most fitting memorial was a new building for the natural sciences at the University of Nebraska, the plans for which were already under way. Dedication of this new building as Bessey Hall was followed later by similar dedications of Bessey halls at Iowa State University and Michigan Agricultural College. Three botany buildings on the campuses of three land-grant universities, the embodiment of "science with practice," were certainly proper tributes to the life of Charles E. Bessey.

# NOTES

## ABBREVIATIONS

*Amer Agr* — *American Agriculturist for the Farm, Garden, and Household*
*AN* — *American Naturalist*
*BP* — *Bessey Papers*
*BG* — *Botanical Gazette*
Division of Botany — Nebraska folder, Corresp. of State Colleges 1894–1909, Division of Botany, USDA, RG 54, National Archives
*Ia Acad Sci* — *Proceedings of the Iowa Academy of Sciences, 1875–1880*
*Ia Agr Soc* — *Report of the Iowa State Agricultural Society*
*Ia Hort Soc* — *Annual Report of the Iowa State Horticultural Society*
*Neb Acad Sci* — *Publications of the Nebraska Academy of Sciences*
*Neb Bd Agr* — *Annual Report of the Nebraska State Board of Agriculture*
*Neb Exp St* — *Annual Report and Bulletins of the Agricultural Experiment Station of Nebraska*
*Neb Hort Soc* — *Annual Report of the Nebraska State Horticultural Society*
Nomenclature, Britton — Letters and Documents on Nomenclature, 1892–98, 1895–1910, Misc. Box, Britton Correspondence
*SPAS* — *Proceedings of the Society for the Promotion of Agricultural Science*

## CHAPTER 1

1. The best introductions to Bessey's life and accounts of his early years are Raymond J. Pool, "A Brief Sketch of the Life and Work of Charles Edwin Bessey," *American Journal of Botany* 2 (December 1915): 505–6 and L. H. Pammel, "Dr. Charles Edwin Bessey," *Prominent Men I Have Met* (Ames, 1928), 3–5; teaching certificate, November 1866, folder of memorabilia, Bessey's Speeches, Articles, and General Botany Notes, Charles Bessey Papers, University of Nebraska Archives, Lincoln (cited as BP); Bessey to I. M. Schmuck, July 22, 1904, Bessey to H. Edwards, December 11, 1906, BP.

2. Frederick Rudolph, *The American College and University: A History* (New York, 1962), 201–40; Laurence R. Veysey, *The Emergence of the American University* (Chicago, 1965), 57–251; Stanley M. Guralnick, "Sources of Misconception on the Role of Science in the Nineteenth-Century American College," *Isis* 65 (September 1974): 352–66; Stanley M. Guralnick, *Science and the Ante-Bellum American College* (Philadelphia, 1975); Charles Weiner, "Science and Higher Education," in David D. Van Tassel and Michael G. Hall (eds.), *Science and Society in the United States* (Homewood, Ill., 1966), 163–89; Lawrence

201

A. Cremin, *American Education: The National Experience, 1783–1876* (New York, 1980); Lawrence A. Cremin, *The Transformation of the School: Progressivism in American Education, 1876–1957* (New York, 1964), 41–50; Lawrence A. Cremin, *American Education: The Metropolitan Experience, 1876–1980* (New York, 1988), 153–55, 242–47, 375–409, 555–65; Robert V. Bruce, *The Launching of Modern American Science, 1846–1876* (Ithaca, 1987), 75–93, 159–65, 287–94, 326–30.

3. Bruce, 326–38; Edward Danforth Eddy, Jr., *Colleges for Our Land and Time: The Land-Grant Idea in American Education* (New York, 1957); Alan I Marcus, *Agricultural Science and the Quest for Legitimacy* (Ames, 1985), 32–42; Earle D. Ross, *Democracy's College: The Land-Grant Movement in the Formative Stage* (Ames, 1942), 1–45; Rudolph, 241–49; Alfred Charles True, *A History of Agricultural Education in the United States, 1785–1925* (Washington, D.C., 1929), 45–83; Veysey, 70–73.

4. True, 100.

5. Ibid., 88–106.

6. Ibid., 106–19; Marcus, 32–42; Rudolph, 253–63.

7. W. J. Beal, *History of the Michigan Agricultural College and Biographical Sketches of Trustees and Professors* (Lansing, 1915), 52–60; Madison Kuhn, *Michigan State: The First Hundred Years* (East Lansing, 1955), 3–71.

8. Kuhn, 58–60.

9. A. S. Welch to Bessey, January 16, 1870, BP; E. A. Bessey to H. Knapp, September 25, 1924, AD-3 Bessey, Pammel Papers, University Archives, Iowa State University Library, Ames.

10. Bessey to L. H. Pammel, April 28, 1898, BP; Pammel, 4–5; Pool, 506–7; Earle D. Ross, *A History of Iowa State College of Agriculture and Mechanic Arts* (Ames, 1942), 1–70.

11. Bessey, "Laying the Foundations," *Annals of Iowa* 9 (1909): 40–44.

12. Ibid.

13. Bessey to A. S. Welch, December 1870, Bessey, "The Work of a Year," a paper to the Eastern Iowa Horticultural Society, December 1870, AD-3 Bessey, Pammel Papers.

14. Bessey, "Laying the Foundations," 27–30.

15. *Fourth Biennial Report of the Board of Trustees of the Iowa State Agricultural College and Farm to the Governor of Iowa* (1871), 6–13; Ross, *Iowa State College,* 140–47; Marcus, 7–42 (Marcus divides the agriculturists into those who looked to science as a model for successful farming and those who looked to business as a model).

16. Bessey, "Laying the Foundations," 32.

17. Bessey letter in Daniel Coit Gilman (ed.), *Recent Information Respecting Agricultural Education Elsewhere* (Berkeley, 1874), 2.

18. Ibid., 2–3; Kuhn, 96.

19. Bessey to I. P. Roberts, April 10, 1899, BP; Pammel, 22; Isaac Phillips Roberts, *Autobiography of a Farm Boy* (Ithaca, 1946), 100.

20. E. Limes to Bessey, January 28, 1871, E. Tupper to Bessey, March 12, 1872, B. F. Gue to Bessey, October 21, 1872, J. W. Shaffer to Bessey, April 8, 1873, November 17, 1874, J. L. Budd to Bessey, June 6, 1873, November 20, 1873, July 5, 1875, J. Mathews to Bessey, September 21, 1874, E. L. Sturtevant to Bessey, 1875, BP; *Fourth Biennial Report,* 82–89; Bessey, "Injurious Insects," *Annual Report of the Iowa State Horticultural Society* 6 (1871): 162–67 (cited as *Ia Hort Soc*); Bessey, "Report on Insects Injurious to the Plants and Animals of the Farm," *Report of the Iowa State Agricultural Society for the Year 1874,* 232–53 (cited as *Ia Agr Soc*).

21. *Fourth Biennial Report;* Roberts, 96–103; Ross, *Iowa State College,* 121–24.

22. *Ia Hort Soc* 9 (1874): 154.

23. Ross, *Iowa State College*, 93–98.
24. *Report of the Joint Committee of the Fifteenth General Assembly of Iowa Appointed to Visit the State Agricultural College and Farm* (Des Moines, 1874); Ross, *Iowa State College*, 96.
25. R. Haigh, Jr., to Bessey, December 20, 1873, A. J. Welch to Bessey, February 2, 1874, BP.
26. *Report of the Joint Committee*, 653–67.
27. Ibid., v–vi; Ross, *Iowa State College*, 95–98.
28. W. H. Wynn to Miss King, February 8, 1906, AD-3 Bessey, Pammel Papers.
29. Ibid.
30. A. Hunter Dupree, *Asa Gray, 1810–1888* (Cambridge, Mass., 1959).
31. A. Gray to Bessey, January 13, May 24, July 18, 1970, May 16, 1973, B.G78, Asa Gray Papers, American Philosophical Society Library, Philadelphia.
32. All by Bessey, "*Lemna polyrrhiza*," *American Naturalist* (cited as AN) 6 (1872): 636; "Sensitive Stamens in Portulaca," ibid. 7 (1873): 464–65; "Double Thalictrum," ibid. 8 (1874): 499; "*Adoxa Moschatellina L.* in Iowa," ibid., 690; "*Botrychium Lunaria Swartz*," ibid., 691.
33. A. Gray to Bessey, March 5, 1870, July 20, 1871, Gray Papers; A. S. Packard, Jr., to Bessey, May 3, 1877, BP; Bessey, "Observations on *Silphium laciniatum*, the So-Called Compass Plant," *AN* 11 (1877): 486–89 and *Proceedings of the Iowa Academy of Sciences, 1875–1880*, 10 (cited as Ia Acad Sci); Bessey, "*Silphium laciniatum*," *AN* 11 (1877): 564; Bessey, "Further Observations upon *Silphium laciniatum*," *Ia Acad Sci*, 13. Bessey's contributions to understanding the compass plant are discussed in Benjamin Alvord, "On the Compass Plant," *AN* 16 (1882): 633–35.
34. S. Watson to Bessey, July 12, 1872, A. Gray to Bessey, August 17, September 13, 1872, Gray Papers; Dupree, 384–85; Pool, 507.
35. Dupree, 332–54.
36. A. Gray to Bessey, September 7, 1872, Gray Papers.
37. Bessey, "Some Personal Remembrances of Asa Gray," Botanical Seminar, November 18–19, 1908, folder 1, General Botany Notes, BP.
38. Gilman, 2.
39. I. P. Roberts to Bessey, December 5, 1874, BP; A. Gray to Bessey, August 21, December 15, 1874, Gray Papers; Pool, 507.
40. Quoted in Pammel, 7.
41. Bessey to J. LeConte, January 17, 1876, LeConte to Bessey, March 3, 1876, BP.
42. Throughout, the term *lower plants* is used in the nineteenth-century sense. At first it referred to the Cryptogams (as opposed to the Phanerogams) when plants generally were divided into these two major groups. As this usage came into disfavor by the late nineteenth century, "lower plants" was used loosely to refer to algae and fungi and sometimes also to mosses and ferns. By the 1880s Bessey divided plants into seven primary divisions, and shortly after the turn of the century he departed from general usage and used fourteen divisions.
43. *Structural physiology* is the term used to refer to this type of study of the relationship of form and function. Bessey, "Some Effects of Low Temperature upon Plants, with a Review of the Nature of Protoplasm," *Ia Hort Soc* 11 (1876): 88–93; Bessey, "A Classification of the Tissues of Plants," *Ia Acad Sci*, 20; Bessey, "The Morphology of the Iris Leaf," ibid., 22; Bessey, "Some Observations on the Action of Frost upon Leaf-Cells," *Proceedings of the AAAS* 31 (1882): 464–65; Pammel, 7; Pool, 507.
44. Bessey, "Microscopic Examinations at the Agricultural College upon Leaf and Cell Growth," *Ia Hort Soc* 8 (1873): 22; Bessey, "Notes on the Colors of the Native Wild

Flowers of Iowa," *Ia Acad Sci,* 8; Bessey, "Some Observations upon the Growth of Plants, Made by Means of the Arc-Indicator," ibid., 10.

45. Ibid.; Bessey, "Notice of a Simple Dendrometer," *Ia Acad Sci,* 24; Bessey, "An Easily Made Observation," *Botanical Gazette* (cited as *BG*) 6 (1881): 172.

46. All by Bessey, "Sensitive Stamens," 464–65; "Notes on the Dimorphism of *Oxalis violacea,*" *Ia Acad Sci,* 11; "Note on the Forms of the Flowers of *Lithospermum longiflorum,*" ibid., 13. The latter he considerably revised and published as "The Supposed Dimorphism of *Lithospermum longiflorum,*" *AN* 14 (1880): 417–21.

47. N. S. Townshend to Bessey, July 6, 1874, BP.

48. C. F. Clarkson to Bessey, June 16, 1874, BP.

49. W. B. Niles to Bessey, September 26, 1871, Bessey to E. F. Smith, December 17, 1901, BP.

50. C. A. White to Bessey, September 26, 1871, BP; Bessey, "Contributions to the Flora of Iowa," *Fourth Biennial Report,* 89–127; Bessey, "Report on Botany, Zoology, and Horticulture," *Fifth Biennial Report, Iowa State Agricultural College* (1873), 91–96.

51. C. A. White to Bessey, September 11, 1871, W. D. Holway to Bessey, July 31, 1883, BP.

52. Bessey, "Report on Botany, Zoology, and Horticulture," 91–96.

53. Bessey, "Contributions to the Bryology of Iowa," *Ia Acad Sci,* 7; Bessey, "A Preliminary Catalogue of the Lichens of Iowa," ibid., 8.

54. Bessey, "On Injurious Fungi," *Sixth Biennial Report, Iowa State Agricultural College* (1875), 128–33 and *Seventh Biennial Report, Iowa State Agricultural College* (1877), 185–204.

55. General impressions drawn from *Ia Hort Soc.*

56. Ibid. 8 (1874): 15.

57. Ibid., 26.

58. A. S. Welch, "The Relation of the Agricultural College to Horticulture," ibid. 9 (1875): 148–60.

59. C. C. Parry, "Botany and Horticulture," ibid. 8 (1874): 56.

60. H. H. McAfee, "Pre-Natal Influences," *Proceedings of the Sixth Annual Meeting of the Eastern Iowa Horticultural Society* (bound with *Ia Hort Soc* 10 [1875]: 349–50).

61. *Ia Hort Soc* 10 (1875): 157, 178–80, 228–29; 12 (1877): 107.

62. J. M. Shaffer to Bessey, April 22, 1873, November 17, 1874, January 25, 1877, C. V. Riley to Bessey, October 22, 1873, BP; Bessey, "Report on Insects Injurious," 232–53; Bessey, "A Preliminary Catalogue of the Orthoptera of Iowa," *Ia Acad Sci,* 8 and in *Seventh Biennial Report,* 205; Bessey, "On the Distribution of the Seventeen Year Cicada of the Brood of 1878 in Iowa," MSS, CX.B470, Iowa State University Library.

63. C. V. Riley to Bessey, March 8, October 22, 1873, March 11, April 20, 1874, February 26, July 24, 1876, March 31, 1880, J. M. Shaffer to Bessey, December 31, 1874, February 5, 1877, B. Sherman to Bessey, October 30, 1876, BP.

64. Bessey, "Fungus on Cottonwood" and "A New Fungus on the Ash," *Ia Hort Soc* 10 (1875): 86–88; "Effects of Low Temperature," 88–93.

65. Ibid. 9 (1874): 6–7, 181–83; 10 (1875): 157–63; 11 (1876): 251–60; 12 (1877): 168–75.

66. S. Calvin to Bessey, May 20, 1874, R. Burgess to Bessey, February 13, 1875, BP.

67. *Ia Acad Sci,* 2–10.

## CHAPTER 2

1. Bessey, "On a Scientific Course of Study," *Aurora* (May 1877), 5–6; (June 1877), 5; (July 1877), 6; *Fourth* and *Fifth Biennial Report;* Ross, *Iowa State College,* 121–24; Roberts, 96–103.

2. Ross, *Iowa State College,* 157–58.

3. Bruce, 75–93, 330–38; Margaret W. Rossiter, *The Emergence of Agricultural Science: Justus Liebig and the Americans, 1840–1880* (New Haven, 1975), 49–124.

4. Bessey, "The Development of Botanical Laboratories in the West," folder 6, Botany General Notes, BP; Pool, 510–11.

5. Bessey, "Choosing a Microscope," *Aurora* (April 1874), 7–8; Bessey, "Using a Microscope," ibid. (May 1874), 7–8; Bessey, "Development of Botanical Laboratories."

6. Bessey, "Greetings to the Class of 1875," folder 6, Botany General Notes, BP.

7. A. S. Welch to Bessey, December 6, 1878, R. J. Beck to Bessey, February 23, 1882, Bessey to L. H. Pammel, April 28, 1898; H. B. Ward, "Bessey Memorial Program," May 14, 1915, Bessey folder, File Drawer 20, BP.

8. L. H. Pammel to Bessey, April 28, 1898, BP; Pammel, 19–22.

9. Ward, "Bessey Memorial Program."

10. T. H. McBride to Bessey, January 6, 1882, E. G. Knight to Bessey, April 15, 1882, J. C. Arthur to Bessey, August 11, 1882, L. J. Martin to Bessey, December 9, 1882, BP.

11. Bessey, *Geography of Iowa, A Supplement to the Eclectic Series of Geographies* (Cincinnati, 1878); Van Antwerp, Bragg and Company to Bessey, December 14, 22, 1877, BP.

12. Holt and Company to Bessey, May 1878, BP.

13. Holt and Company to Bessey, n.d.[1874–5], May 1878, BP.

14. Handwritten answer on margin of H. Holt to Bessey, May 20, 28, 1878, contract with Holt and Company, June 1, 1878, Bessey to Holt, June 1, 1878, BP.

15. Bessey to A. Gray, May 31, 1878, BP; Gray to Bessey, June 4, 1878, B.G78, Gray Papers.

16. All by Bessey, "Botanical Aspects of Apple Blight," *Eighth Biennial Report, Iowa State Agricultural College* (1879), 106–7; "Synopsis of a Lecture Upon the Leaf: Its Structure, Functions and Climatic Modification," *Ia Hort Soc* 14 (1879): 130–34; "The Flower: Its Structural and Functional Meaning," ibid. 15 (1880): 174–82; "On the Affinities of the Uredineae," *Ia Acad Sci,* 14; "On a Botanical Map of the United States," ibid., 18; "Sketch of a Natural Arrangement of Plants," ibid.; "Sketch of a Course of Laboratory Practice in Higher Botany," ibid., 20; "Classification of Tissues;" "Morphology of Iris Leaf;" "A Genealogical Tree of the Vegetable Kingdom," ibid., 24.

17. Bessey to Holt and Company, February 1, September 17, December 15, 1879, January 2, 13, April 12, 19, May 10, 1880, BP. There are numerous other letters on this subject to Holt and Company, but most concern illustrations.

18. Bessey, *Botany for High Schools and Colleges* (Philadelphia, 1880), iv–v; Bessey to Holt and Company, May 10, 1880, BP.

19. Emanuel D. Rudolph, "The Botanical Textbook in Nineteenth-Century America As a Reflection of Botanical and Cultural Trends," paper presented to the Midwest Junto for the History of Science, 1975.

20. The chapter titles of Part I illustrate this emphasis: "Protoplasm," "The Plant Cell," "The Cell Wall," "The Formation of New Cells," "The Products of the Cell," "Tissues," "The Tissue Systems," "Intercellular Spaces and Secretion Reservoirs," "The Plant Body," "The Chemical Constituents of Plants," "The Chemical Processes in the Plant," and "The Relations of Plants to External Agents." Bessey, *Botany,* iv.

21. G. L. Goodale to Bessey, April 11, 1874, BP; Bessey, *Botany;* A. G. Morton, *History of Botanical Science* (London, 1981), 381-440.
22. Asa Gray, *Gray's School and Field Botany,* rev. ed. (New York, 1887); Dupree, 169-70, 202-03.
23. *BG* 4 (1879): 235; Dupree, 48-51, 169-70, 202.
24. Asa Gray, *Gray's Lessons in Botany: The Elements of Botany for Beginners and for Schools* (New York, 1887), 10.
25. Bessey, *Botany,* 4th ed. (1885), 202-04.
26. E. W. Hilgard to Bessey, November 27, 1880, BP.
27. J. C. Arthur to Bessey, September 28, 1880, BP.
28. V. M. Spalding to Bessey, August 14, 1880, N. H. Winchell to Bessey, August 30, 1880, S. N. Townshend to Bessey, September 7, 1880, E. A. Smith to Bessey, October 8, 1880, A. N. Prentiss to Bessey, October 29, 1880, D. S. Jordan to Bessey, November 1, 1880, E. W. Hilgard to Bessey, November 27, 1880, A. Winchell quoted in H. Holt to Bessey, November 27, 1880, S. Calvin to Bessey, December 17, 1880, J. T. Rothrock to Bessey, February 20, 1881, BP.
29. P. H. Mell, Jr., to Bessey, October 12, 1880, T. C. Porter to Bessey, December 17, 1880, BP.
30. J. M. Coulter to Bessey, August 24, 1880, BP.
31. W. G. Farlow to Bessey, September 7, 1880, BP.
32. A. Gray to Bessey, August 12, 1880, BP.
33. Dupree, 393-94.
34. E. W. Hilgard to Bessey, November 20, 1880, BP.
35. *American Journal of Science and Arts,* 3rd series, 20 (July-December 1880): 337.
36. A. S. Packard, Jr., to Bessey, August 10, 1880, BP; *AN* 14 (1881): 796-97.
37. J. M. Coulter to Bessey, August 24, 1880, BP.
38. *BG* 5 (1880): 96-98.
39. William Ramsey McNab, *Botany: Outlines of Morphology, Physiology and Classification of Plants,* Specially Revised for American Students by Charles E. Bessey (New York, 1881).
40. Bessey to Holt and Company, October 1, 29, 1878, n.d.[November-December 1878], February 17, 1879, W. R. McNab to Bessey, March 24, 1879, BP; *AN* 15 (1881): 898.
41. Ibid.; Bessey to Holt and Company, November 27, 1880, January 22, 1881, Statement from Holt and Company, June 30, 1882, June 30, 1883, BP.
42. Bessey, *Botany,* 4th ed. (1885), vi; Bessey to Holt and Company, November 27, 1880, May 27, 1881, Holt to Bessey, March 17, May 17, 1881, BP.
43. Bessey to H. Holt, January 9, 1882, BP.
44. Holt and Company to Bessey, January 8, 1880, BP.
45. *AN* 16 (1882): 315; Bessey, "Modern Botany and Mr. Darwin," ibid., 507; ibid. 18 (1884): 534.
46. Holt and Company to Bessey, January 8, 1880, BP.
47. Bessey to Holt and Company, October 29, 1878, Holt to Bessey, January 12, 1884, BP; *AN* 18 (1884): 186-87.
48. Holt and Company to Bessey, September 28, 1885; Bessey, *The Essentials of Botany,* The American Science Series, Briefer Course, vol. V (New York, 1884); *BG* 9 (1884): 184.
49. A. Gray to Bessey, October 10, 1884, Gray Papers; *AN* 18 (1884): 1247-48.
50. *AN* 15 (1881): 999; all by Bessey, "The Superabundance of Pollen in Indian Corn," ibid., 1000; "Remarkable Fall of Pine Pollen," ibid. 17 (1883): 658; "Hybridism in Spirogyra," ibid. 18 (1884): 67-68; ibid., 311; "Glands on a Grass," ibid., 420-21; "Sexual-

ity in Zygnemaceae," ibid., 421–22; "Structure of the Fruit of Porcupine Grass," ibid., 930–31; "Adventitious Inflorescence of *Cuscuta glomerata,*" ibid., 1147; "Curvature of the Stems of Conifers," BG, 9 (1884): 156; "Mode of Opening of the Flowers of *Desmodium sessilifolium,*" ibid.; "A Point in the Structure of the Sterile Flowers of Silphium," ibid., 159; "The Functions of the Leaf," ibid. 18 (1883): 195–200; "Observations of Frost," 464–65; "The Flower," 174–82; "Morphology of Iris Leaf," 22.

51. Bessey, "The Climatic Adaptation of Plants," *Ia Hort Soc* 12 (1877): 97.
52. Bessey, "A Case of Natural Selection," *Ia Acad Sci,* 10.
53. Bessey, "Botany and Mr. Darwin," 507–8.
54. Bessey, "The Flower," 174–82; Bessey, *"Cuscuta glomerata,"* 1145–47.
55. Bessey, *Botany,* 4th ed. (1885), 202–04.
56. *AN* 15 (1881): 44, 53, 132, 227, 473; 16 (1882): 44, 731; 17 (1883): 1066; 18 (1884): 1026.
57. Bessey, "Preliminary List of Cryptogams," *College Bulletin* (1884), 133–50.
58. J. C. Arthur, "Preliminary List of Iowa Uredineae," ibid., 151; *BG* 4 (1879): 115; 7 (1882): 127; 8 (1883): 271; William G. Farlow, "Notes on Some Ustilagineae of the U. S.," ibid., 271–77; Farlow, "Additions to the Peronosporeae of the U. S.," ibid., 37.
59. Bessey, "Insect-Destroying Fungi," *AN* 15 (1881): 52–53; Bessey, "The Fungi Which Produce Mildew on Cotton Goods," ibid., 132; ibid. 16 (1882): 394, 586; Bessey, "A New Species of Insect-Destroying Fungus," ibid. 17 (1883): 1280–81.
60. Bessey, "The Diseases of Plants," *Ia Hort Soc* 17 (1882): 85–86.
61. Bessey, "Apple Blight," 106–7; Bessey, "The Bearing of Plowright's Discovery As to the Germination of Rust-Spores upon the Problem of Wheat-Growing in the North-West," *SPAS* (1883), 31; Bessey, "Two Parasitic Fungi," *Proceedings of the Eastern Iowa Horticultural Society,* bound with *Ia Hort Soc* 18 (1883): 521–23; Bessey, "Popular Descriptions of Some Harmful Plants," *College Bulletin* (1884), 110–32; Bessey, "Corn Smut," *Students' Farm Journal* (Ames) 1 (November 1, 1884), AD-3 Bessey, Pammel Papers.
62. Bessey, "Diseases of Plants," 85–98; Bessey, "On Parasitic and Other Fungi," ibid., 280–84.
63. Ibid.; Bessey, "On Injurious Fungi," 185–204; Bessey, "Injurious Fungi, in Their Relation to the Diseases of Plants," *Proceedings of the American Pomological Society* (1885), 35–43.
64. Andrew Denny Rodgers III, *Erwin Frink Smith: A Story of North American Plant Pathology* (Philadelphia, 1952), 91–118; Kenneth Baker, "Fire Blight of Pome Fruits: The Genesis of the Concept That Bacteria Can Be Pathogenic to Plants," *Hilgardia,* no. 18, 40 (July 1971): 6–33.
65. Bessey, "Injurious Fungi," 38; T. Burrill to Bessey, June 5, November 25, December 23, 1882, BP.
66. *Ia Hort Soc* 11 (1876): 7.
67. Bessey to Iowa Eastern Horticultural Society and an essay (no title), ibid. 10 (1875): 317–19.
68. Bessey, "Climatic Adaptation," 86–97.
69. Ibid., 97.
70. *Ia Hort Soc* 12 (1877): 106–7; 13 (1878): 24, 219–20, 461; 14 (1879): 10, 122–24, 283; 15 (1880): 13, 91–92, 270; 16 (1881): 13, 80; 17 (1882): 11; 18 (1883): 83–84; Bessey, "Botany in Its Relation to Horticulture," ibid. 13 (1878): 222–28; C. C. Parry, "Notes on Rocky Mountain Conifers As Adapted to Cultivation in the Central Northwestern States," ibid., 343–47; J. L. Budd, "Experimental Horticulture," ibid., 217–21; Bessey, "The Condition of the Living Plant in Winter," ibid. 19 (1884): 200–6; Bessey, "Synopsis of Leaf,"

130–34; Bessey, "The Flower," 174–82; Bessey, "Diseases of Plants," 85–98; Bessey, "On Parasitic and Other Fungi," 280–84; Bessey, "Functions of Leaf," 195–200. See Secretary's Annual Reports, 1878–84, *Ia Hort Soc.* A major part of each annual meeting was devoted to the introduction of new varieties of fruits; also see ibid. 13 (1878): 461; 14 (1879): 124–26, 283, 536; 18 (1883): 83–84.

71. Ibid. 14 (1879): 209–10, 421–51.

72. A. S. Welch, "The Effect of Horticulture on the Horticulturist," ibid. 17 (1882): 153–60.

73. Seaman A. Knapp, "The Agricultural College, As It Is, And As It Ought to Be," *Ia Agr Soc* (1881), 243–53.

74. Ibid., 253.

75. Ibid., 255–58; Marcus, 172–76; Joseph Cannon Bailey, *Seaman A. Knapp: Schoolmaster of American Agriculture* (New York, 1945), 73–96.

76. W. G. LeDuc to Bessey, January 14, 1878, G. Thurber to Bessey, February 14, 1881, Bessey to J. A. Garfield, February 17, 1881, D. C. Eaton to Bessey, June 6, 1881, BP; Bessey, "A Government Duty," *AN* 17 (1884): 543; ibid. 18 (1884): 1265.

77. *Proceedings of a Convention of Agriculturists Held in the Department of Agriculture, January 10–18, 1882,* USDA, Report no. 22, 3–39.

78. Circular letter from W. J. Beal, G. O. Caldwell, and E. L. Sturtevant, September 1879, Beal to Bessey, September 13, 1882, BP; *SPAS* (1881), 9–11; W. J. Beal, "The Society for the Promotion of Agricultural Science," ibid. (1907), 27–30.

79. Circular letter from W. J. Beal, G. C. Caldwell, and E. L. Sturtevant, September 1879, Circular letters from Beal, September 29, October 7, November 20, 1879, Beal to Bessey, March 7, 1880, December 23, 1881, L. Sturtevant to Bessey, March 29, 1880, BP; *SPAS* (1881), 9–11.

80. Ibid. (1881), 9–11, 57–58; (1884), 2–7.

81. W. J. Beal to Bessey, September 13, 1882, G. C. Caldwell to Bessey, June 11, 1883, BP; Bessey, "Plowright's Discovery," 31; *AN* 18 (1884): 291, 573, 611, 1072, 1265.

82. E. L. Sturtevant to Bessey, 1875, B. D. Halsted to Bessey, October 25, 1881, A. R. Crandell to Bessey, February 18, 1882, H. N. Parker to Bessey, February 27, 1882, BP; Bessey, "The Plague of Cut-Worms," reprinted from the *Chicago Herald* in the *Ia Hort Soc* 16 (1882): 332–33; Ross, *Iowa State College,* 164–66.

83. H. W. Parker to Bessey, June 7, 1879, Bessey to G. H. Wright, December 29, 1879, C. W. von Codler to Bessey, January 4, 1881, BP; Bessey, "Laying the Foundations," 38; Copy from Minute Book C-1882–Board of Trustees, July 5, 1882, AD-3 Bessey, Pammel Papers; "C. E. Bessey; Our First Acting President," *Alumnus of Iowa State,* no. 3, 29 (October 1933), ibid.

84. H. W. Parker to Bessey, June 7, 1879, C. H. Fernald to Bessey, November 1, 1884, G. H. Wright to Bessey, December 2, 1885, BP.

85. *Ia Agr Soc* (1881), 243–58; Ross, *Iowa State College,* 101–2, 122–26.

86. A. S. Welch, "Science with Practice in Education," *Ia Hort Soc* 16 (1881): 153–60.

87. Ibid., 160; Ross, *Iowa State College,* 98.

88. Copy from Minute Book C-1882–Board of Trustees, July 5, 1882, AD-3 Bessey, Pammel Papers; Ross, *Iowa State College,* 99–101.

89. Ibid.

90. *Iowa State Register* (Des Moines), November 27, 1883, reprinted articles from *Republican* (Cedar Rapids), *Herald* (Carroll), and *Argus* (Webster City).

91. Bessey, "Laying the Foundations," 32.

92. *Iowa State Register,* December 1, 7, 27, 1883; Ross, *Iowa State College,* 99–101.

93. H. G. Grattan to Bessey, October 26, 1884, BP; *Iowa State Register,* December 12, 1883; Ross, *Iowa State College,* 101–04.

94. Bessey to L. H. Pammel, April 28, 1898, BP; Pool, 508.

## CHAPTER 3

1. W. W. Jones to Bessey, September 26, 1884, H. S. Grant to Bessey, October 6, 1884, R. W. Furnas to Bessey, January 25, April 4, 1885, BP.

2. Bessey, "Science and Practice," *Daily Nebraska State Journal* (Lincoln), September 17, 1884.

3. *Annual Report of the Agricultural Experiment Station of Nebraska* (1888), 1–2 (cited as *Neb Exp St*).

4. J. G. Lemmon to Bessey, March 25, 1884, May 4, November 7, 1885, G. B. Loring to Bessey, January 5, 1885, A. H. Curtiss to Bessey, April 2, 27, 1885, M. A. Cerill to Bessey, May 25, July 22, 1885, O. Wilcoxen to Bessey, February 6, 1886, May 14, 1887, H. G. Wilkinson to Bessey, June 16, 1886, K. Whirted to Bessey, September 9, 1886, A. H. Van Fleet to Bessey, August 24, 1887, C. L. Nettleton to Bessey, February 4, 1888, E. C. Blake to Bessey, February 1, 16, 1888, BP; *BG* 10 (1885): 326; 11 (1886): 159.

5. Raymond J. Pool, "The Evolution and Differentiation of Laboratory Teaching in the Botanical Sciences," *Iowa State Journal of Science* 3 (January 1935): 238–39.

6. *AN* 19 (1885): 695; Bessey, "The Movement of Protoplasm in the Styles of Indian Corn," ibid., 888; Bessey, "The Roughness of Certain Uredospores," ibid. 20 (1886): 1053; Bessey, "Botanical Manuals for Students," ibid. 21 (1887): 376–79; John M. Coulter, "Laboratory Appliances," *BG* 10 (1885): 409; ibid. 11 (1886): 20; Bessey, "Specimens and Specimen Making," ibid., 132; Bessey, "Herbarium Cases," ibid., 186–87; Bessey, "Books of Reference," ibid., 193; J. C. Arthur, "Some Botanical Laboratories of the United States," ibid. 10 (1885): 395–406.

7. Thomas R. Walsh, "Charles E. Bessey and the Transformation of the Industrial College," *Nebraska History* 52 (Winter 1971): 383–409; Thomas R. Walsh, "Charles E. Bessey: Land-Grant College Professor," Ph.D. diss., University of Nebraska–Lincoln, 1972; Robert N. Manley, *Centennial History of the University of Nebraska: I. Frontier History, (1869–1919)* (Lincoln, 1969), 34–39, 61–62, 100–3.

8. Manley, 102–3; Robert Platt Crawford, *These Fifty Years: A History of the College of Agriculture of the University of Nebraska* (Lincoln, 1925), 46–48, 56–57; Walsh, "College Professor."

9. Bessey, "Science and Practice"; *Neb Exp St* (1888), 1–2; Catalogue of 1884–85, University of Nebraska Archives; Crawford, 50–51, 57–58; Manley, 103; Walsh, "College Professor"; *American Agriculturist for the Farm, Garden, and Household* 47 (1888): 14 (cited as *Amer Agr*).

10. Crawford, 57–61; Walsh, "College Professor."

11. Ibid.

12. Ibid.

13. Crawford, 33–37; Manley, 108; Walsh, "College Professor."

14. R. W. Furnas to Bessey, August 17, October 13, 1885, February 8, 1886, Bessey to Furnas, September 16, 1891, September 5, 17, 1892, BP; *Amer Agr* 47 (1888): 14; Bessey, "The Grasses and Forage Plants of Nebraska," *Annual Report of the Nebraska State Board of Agriculture* (1886), 204–37 (cited as *Neb Bd Agr*); Bessey and Herbert J. Webber, "The Grasses and Forage Plants, and the Catalogue of Plants," ibid. (1889), 144–74;

Bessey, "A Preliminary List of the Grasses of Nebraska," ibid. (1891), 124–30; Bessey, "A Preliminary Description of the Native and Introduced Grasses of Nebraska," ibid. (1892), 209–79; Walsh, "College Professor."

    15. S. Bernard to Bessey, February 4, 1886, November 23, December 1, 1887, BP; Bessey, "On the Fertilization, Crossing, and Hybridization of Plants," *Annual Report of the Nebraska State Horticultural Society* (1891), 100–13 (cited as *Neb Hort Soc*); Bessey, "A Second Report upon the Native Trees and Shrubs of Nebraska," ibid. (1892), 154–85; Bessey, "List of Nectar-Producing Plants," ibid. (1887–88), 203–09.

    16. *Breeder's Gazette* (1887), 251, (1888), 212–13, (1889), 230, (1890), 8, 152.

    17. F. M. Hexamer to Bessey, March 6, 1891, W. R. Goodwin, Jr., to Bessey, October 14, 1891, BP. The following were all by Bessey in *Amer Agr*: "Grasses for the Nebraska Plains," 45 (1886): 400; "Clover upon the Nebraska Plains," 443; "How Nebraska Conducted Her State Fair," 472; "Destruction of Rocky Mountain Forests," 520; "Tree Planting on the Plains," 46 (1887): 9; "Trees and Blizzards," 47; "Horticulture in Nebraska," 99; "Ornamental Wild Flowers of the Plains," 156; "Some Western Weeds," 199; "The Country School and the Farmer's Boy," 243; "Arbor Day in Nebraska," 327; "A Good Western Grass," 367; "Two Promising Native Cherries," 418; "Agricultural Progress in Nebraska," 481; "A Model Agricultural Fair," 551; "Elementary Agricultural Schools," 47 (1888): 10; "What One Farmers' Club Did," 49; "Horticulture on the Plains," 98; "Some Useful Nebraska Grasses," 49 (1890): 635; "The Improvement of Wild Grasses," 51 (1892): 356. The following items were all by Bessey in *Breeder's Gazette:* "Forage Plants," (1886), 381; "The Relations of Vegetation to Stock Growing," (1887), 251; "Fine Grasses for Fine Stock," (1888), 212–13; "Abuse of Cold Storage," (1890), 8; "The Native Grasses of the Plains," 171; "Grasses of the West Again," 252; "Wild or Tame Grasses for the Plains?" 375; "The Meadow in Winter," 484–85; "Forage Plants on the Plains," (1891), 468.

    18. Crawford, 49–50.

    19. Alfred Charles True, *A History of Agricultural Experimentation and Research in the United States, 1607–1925,* Misc. Publication No. 251, USDA (Washington, D.C., 1937), 49–118; Marcus, 59–126; Charles E. Rosenberg, "The Adams Act: Politics and the Cause of Scientific Research," *Agricultural History* 38 (January 1964): 3–12; Rosenberg, "Science, Technology, and Economic Growth: The Case of the Agricultural Experiment Station Scientist, 1875–1914," ibid. 44 (January 1971): 1–20; Rosenberg, "Science and Social Values in Nineteenth-Century America: A Case Study in the Growth of Scientific Institutions," *No Other Gods: On Science and American Social Thought* (Baltimore, 1978), 135–52.

    20. Bessey, "The Demands Made by Agriculture upon the Science of Botany," *SPAS* (1885), 16–18; Bessey, "A Duty Which We Owe to Science," ibid. (1887), 28–30.

    21. *AN* 19 (1885): 172.

    22. *Neb Exp St* (1888), 5–21; Bessey, "Science and Practice."

    23. Bessey, "Demands by Agriculture," 16–18; G. P. Clinton, "Botany in Relation to Agriculture," *Science* 43 (1916): 1–13.

    24. Report to the Regents, December 1884, BP; Bessey, "Science and Practice."

    25. R. W. Furnas to Bessey, March 12, 1886, BP; Crawford, 51.

    26. F. L. Scribner to Bessey, January 18, February 17, 1886, March 8, 1887, G. W. E. Dorsey to Bessey, January 5, 1886, J. Laird to Bessey, February 10, 1886, C. F. Manderson to Bessey, January 17, 1887, BP; *Amer Agr* 48 (1889): 144.

    27. *Amer Agr* 48 (1889): 144; A. Hunter Dupree, *Science in the Federal Government: A History of Policies and Activities to 1940* (New York, 1957), 161–69.

    28. Bessey, "Demands by Agriculture," 16–18; *AN* 20 (1886): 65–66, 448–50; Marcus, 89–126.

29. *Report of the Commissioner of Agriculture,* USDA (1885-87); Bailey, 98-101; Marcus, 127-32, 161-211; Ross, *Iowa State College,* 163; True, *Agricultural Experimentation,* 118-30.

30. G. Dorsey to Bessey, February 23, 1887, C. F. Manderson to Bessey, January 23, 1888, BP; Marcus, 184-216; Ross, 163. There is nothing specific in the Bessey Papers to show anything Bessey wrote regarding the Hatch Act. In his Papers, however, there is a printed copy of the Senate bill on which part of Section 2 is checked and underlined; Box of Bessey Speeches, Articles, General Botany, BP.

31. *Neb Exp St* (1888), 20-22. Bessey presented a paper on his proposals at the 1890 meeting of the Association of American Agricultural Colleges and Experiment Stations; *AN* 24 (1890): 102.

32. *Neb Bd Agr* (1890), 38.

33. *Amer Agr* 48 (1889): 84.

34. Ibid., 324, 371.

35. *Neb Exp St* (1888), 1-22; (1889), 259-353; *Amer Agr* 46 (1887): 378.

36. *Neb Exp St* (1889), 295-353.

37. Richard A. Overfield, "Hog Cholera, Texas Fever, and Frank S. Billings: An Episode in Nebraska Veterinary Science," *Nebraska History* 57 (Spring 1976): 99-128.

38. Walsh, "College Professor."

39. Crawford, 60-61.

40. Ibid., 64-66. There are numerous references in the Bessey Papers to his participation in the Farmers' Institutes.

41. Manley, 108-10.

42. *Neb Exp St* (1888), 18-19.

43. Manley, 110.

44. Marcus Benjamin, "Charles Edwin Bessey," *University Journal* 8 (January 1912): 66.

45. Bessey, "Education for the Farmer's Boy," *Nebraska Farmer* 13 (1889): 814, 847, 871, 883, 910; Bessey, "The Diseases of Farm and Garden Crops," ibid. 14 (1890): 89, 129-30, 151, 165, 189, 209, 250, 293, 333.

46. G. D. Swezey to Bessey, July 4, 1889, Bessey to Chancellor, November 25, 1891, BP; Walsh, "College Professor."

47. Manley, 108-9.

48. Bessey to H. Holt, August 1, 1891, BP; Crawford, 70-71; Manley, 119-25, 141-42, 175-76.

49. Bessey to H. Holt, January 29, 1892, BP.

50. Bessey, *Elementary Botanical Exercises for Public Schools and Private Study* (Lincoln, 1892); Bessey to Blake, November 14, 1892, Bessey to J. O. Taylor, April 26, 1892, Bessey to J. J. Campbell, March 15, 1893, Bessey to J. W. Kerces, September 17, 1894, American Book Company to Bessey, December 15, 1894, May 27, 1895, Bessey to Holt and Company, June 26, 1895, Bessey to American Book Company, June 28, 1895, BP; *BG* 17 (1892): 264.

51. J. Gardiner to Bessey, January 3, 1889, July 24, 1892, H. Holt to Bessey, October 31, 1889, July 17, 1891, Bessey to Holt, August 1, 1891, Bessey to L. Pammel, December 4, 1894, Bessey to L. A. Osborne, January 17, 1895, Bessey to Chancellor, November 4, 1898, BP; *BG* 18 (1893): 325.

52. All by Bessey, "The Question of Bisexuality in the Pond-Scums," *BG* 10 (1885): 334; "Further Observations on the Adventitious Inflorescence of *Cuscuta glomerata,*" ibid.; "The Opening of the Flowers of *Desmodium sessilifolium,*" *AN* 19 (1885): 711-13; "Attempted Hybridization Between Pond-Scums of Different Genera," ibid., 800-2; "The

Growth of *Tulostoma mammosum*," ibid. 21 (1887): 665-66; "An Overlooked Function of Many Fruits," ibid. 22 (1888): 531; "Two Big-Rooted Plants of the Plains," ibid. 23 (1889): 174-76.

53. W. A. Henry to Bessey, September 30, 1889, Bessey to J. Michels, October 30, 1891, Bessey to F. S. Billings, August 29, 1891, BP; Bessey, "Diseases of Farm and Garden," 89, 129-30, 151, 165, 189, 209, 250, 293, 333; Bessey, "The Smut of Indian Corn," *Neb Exp St,* Bulletin #11 (1889), 295-305; Bessey, "Wheat Rust," ibid., Press Bulletin #2.

54. Bessey to Director [H. H. Nicholson], February 6, August 25, 1892, Bessey to L. H. Pammel, September 10, 1892 [copies of same letter to fifteen other persons], BP.

55. C. H. Peck to Bessey, May 15, August 31, 1885, W. Farlow to Bessey, February 16, 1887, BP; all by Bessey, "Injurious Fungi," 35-43; "The Study of Parasitic Fungi," *AN* 19 (1885): 170-71; "The Abundance of Ash Rust," ibid., 886-87; "Pear Blight Bacteria and the Horticulturists," ibid. 20 (1886): 166; "The Rust of the Ash Tree," ibid., 806; "The Study of Plant Diseases," ibid. 21 (1887): 276-77; "The Host-Index of the Fungi of the United States," ibid. 24 (1890): 1196-97; "Some Bad Station Botany," ibid., 1197; "Wheat Smut," ibid., 1197-98.

56. Bessey, "Injurious Fungi," 35-43; Bessey, "A New Work on 'Plant Morphology,' " *AN* 24 (1890): 1082-83.

57. *AN* 18 (1884): 291, 573-77, 611, 614, 1265; 19 (1885): 172, 444, 713; 21 (1887): 49.

58. Bessey, "Some Useful Nebraska Grasses," 635; Bessey, "Native Grasses of the Plains," 171; Bessey, "Wild or Tame Grasses," 375; Bessey, "Forage Plants on the Plains," 468; Bessey, "Preliminary List of Grasses," 124-30; Bessey, "Preliminary Description of Native and Introduced Grasses," 209-79; Bessey, "A Dozen Grasses and Clovers for Nebraska," *Neb Bd Agr* (1890), 100-8; Per Axel Rydberg, "Some Grasses of Southwestern Nebraska," ibid. (1891), 134-37; J. M. Bates, "The Grasses of Northwestern Nebraska," ibid., 130-34; Jared Smith, "The Grasses of the Sand Hills of Northern Nebraska," ibid., 280-91; Albert Woods, "A Few Notes on the Grasses of the 'Bad Lands' [of Nebraska]," ibid. (1892), 291-93.

59. All by Bessey, "Plant Migrations," *AN* 19 (1885): 398-99; "The Grass Flora of the Nebraska Plains," ibid. 22 (1888): 171-72; "Grasses for the Nebraska Plains," 400; "The Grass Problem in Nebraska," *SPAS* (1889), 17-19; "The Forage Problem on the Plains," ibid. (1890), 17-20.

60. Bessey, "Forage Plants," 381; Bessey, "Grasses and Forage Plants," 204-37; Bessey and Herbert J. Webber, "Grasses and Forage Plants and Catalogue," 144-74; Bessey, "Forage Problems on the Plains," 17-20; Bessey, "Improvement of Wild Grasses," 356.

61. Governor to Bessey, January 2, 1893, Bessey to L. O. Howard, July 25, 1893, BP; Bessey, "The Native and Introduced Forage Plants of Nebraska," [abstract in July 1893, BP].

62. Bessey to Chancellor and Regents, December 4, 1897, BP; Bessey, "Grasses and Other Forage Plants," *Neb Bd Agr* (1896), 80-93; Bessey, "The Forage Problem in Eastern Nebraska, Central Nebraska, and on the High Plains," ibid. (1897), 111-20; Bessey, "The Grasses of Nebraska," ibid. (1904), 175-88; Bessey, "Observations on Buffalo Grass," *SPAS* (1899), 105-6.

63. Bessey to L. H. Dewey, September 11, 1894, Nebraska folder, Correspondence of State Colleges, 1894-1909, Division of Botany, USDA, RG 54, National Archives (cited as Division of Botany); Bessey to C. Piperin, September 15, 1897, BP; Bessey, "The Russian Thistle in Nebraska," *Neb Exp St,* Bulletin #31 (1893), 67-77; Bessey, "The Weeds of Nebraska," *SPAS* (1893), 33-43; Bessey, "The Passing of the Russian Thistle," ibid. (1899), 83-85.

64. C. Callahan to Bessey, November 25, December 9, 1895, F. H. Newell to Bessey, June 1, 1896, Bessey to Chancellor and Regents, December 4, 1897, Report of the Botanist to the Experiment Station, [December 1898], BP; Bessey and Albert Woods, "Transpiration; or, the Loss of Water from Plants," paper to AAAS and in *Contributions of the Botany Department* (1892); Bessey, "Some Facts in Vegetable Physiology Related to Problems in Irrigation," *Neb Bd Agr* (1894), 128–32.

65. All by Bessey, "Fertilization, Crossing, and Hybridization," 100–13; "The Botany of the Apple Tree," *Neb Hort Soc* (1894), 7–36; "Botany of the Grape," ibid. (1895), 7–26; "Notes on the Botany of the Strawberry," ibid. (1896), 237–40; "Some Facts in Plant Physiology Bearing upon Horticultural Practices," ibid. (1898), 135–45; "Physiology of the Apple Tree," ibid. (1899), 27–45; "Crop Improvement by Utilizing Wild Species," ibid. (1906), 116–23; "The Physiology of Pruning," ibid. (1908), 94–98; "A Preliminary List of Honey-Producing Plants of Nebraska," *Neb Exp St,* Bulletin #40 (1895), 141–52; "Old World Contributions to Western Orchards," *SPAS* (1901), 26–34.

66. R. W. Furnas to Bessey, March 9, 1896, February 12, 1900, Bessey to Furnas, October 17, 1896, Bessey to Chancellor and Regents, December 4, 1897, Report for the Department of Botany, [December 1896], BP.

67. Bessey, "Report of the Botanist: A Preliminary Account of Diseases of the Farm Crops of Nebraska," *Neb Bd Agr* (1898), 139–61; Bessey, "Notes on the Apple Scab," *Neb Hort Soc* (1901), 100–8; Bessey, "Preliminary Paper on Diseases of Grapes in Nebraska," ibid. (1903), 86–89.

68. Report for the Department of Botany, [December 1896], Bessey to Chancellor and Regents, December 4, 1897, Report of the Botanist to the Experiment Station, [December 1898], BP.

69. Bessey to Chancellor Canfield, August 17, 1891, Bessey to T. L. Lyon, July 11, 1900, J. T. Stack to Bessey, December 26, 1900, BP.

70. Bessey to Chancellor Canfield, August 17, 1891, Bessey to B. Halsted and others, November 1892, Bessey to Halsted, December 24, 1897, September 7, October 2, 4, 1900, October 9, 1901, Bessey to A. F. Woods, November 24, 1900, Bessey to C. R. Barnard, December 11, 1901, BP; Bessey, "The Homologies of the Uredineae (The Rusts)," *AN* 28 (1894): 989–96.

71. Bessey to B. Galloway, October 21, 1897, Bessey to Director of Experiment Station, January 8, 1898, Report of the Botanist to the Experiment Station, [December 1898], BP.

72. Bessey to Professor Wilson and others, May 10, 1898, Bessey to G. W. Bear, May 25, 1898, V. K. Chestnut to Bessey, July 23, August 2, 1898, Bessey to Editor of *Nebraska Farmer,* May 31, 1899, W. Alexander to Head of Experiment Station, May 26, 1900, BP; Bessey to L. H. Dewey, November 23, 1898, Division of Botany.

73. Bessey to C. R. Barnard, December 11, 1901, Bessey to E. A. Burnett, May 6, 1904, Bessey to J. Hamilton, May 17, 1904, BP.

74. F. L. Scribner to Bessey, March 13, 21, 1895, BP; Bessey to Scribner, March 16, 1895, Scribner Papers, Manuscripts Division, Library of Congress; Bessey to F. Coville, January 16, 1894, Division of Botany.

75. Bessey to H. L. Shantz, September 7, 1907, BP.

76. H. J. Webber to Bessey, September 5, 1892, November 12, 1894, P. J. O'Gara to Bessey, December 4, 1902, January 17, 1903, H. Metcalf to Bessey, May 9, 1904, E. Field to Bessey, December 5, 1909, BP; Pammel, 10–11; Paul B. Sears, "Botanists and the Conservation of Natural Resources," in William Campbell Steere (ed.), *Fifty Years of Botany* (New York, 1958), 360–61.

77. Bessey to W. O. Atwater, February 18, 1890, Nebraska 1888–1901, State Corresp.

1888–1937, Experiment Stations, USDA, RG 54, National Archives; F. L. Scribner to Bessey, June 28, August 3, 1885, F. C. Nesbit to Bessey, March 27, 1886, W. B. Barrows to Bessey, August 25, 1886, N. Colman to Bessey, April 18, 1887, B. E. Fernow to Bessey, November 26, 1888, F. V. Coville to Bessey, July 29, 1889, Bessey to E. Willits, May 16, November 2, 1892, BP.

78. Bessey to F. Coville, May 21, August 9, 1894, C. L. Shear to Coville, February 12, 1895, Division of Botany; Bessey to F. L. Scribner, June 8, 1895, June 3, 1897, Bessey to J. Wilson, November 1, 1897, Bessey to H. P. McIntosh, November 4, 1897, Bessey to S. Harris, November 4, 1897, Wilson to Bessey, November 4, 1897, Bessey to D. H. Mercer, January 14, 1898, Bessey to W. E. Moe, May 10, 1899, Bessey to O. F. Cook, May 27, 1899, BP.

79. Bessey to C. F. Miller, May 23, 1899, BP; Crawford, 7–83.

80. Ibid.

81. Bessey to S. T. Maynard, November 1, 1898, BP.

82. Bessey to E. G. Stephens, June 14, 1899, BP.

83. Manley, 141–43, 175–77.

## CHAPTER 4

1. There are numerous references to the "new botany" in the *American Naturalist* and the *Botanical Gazette*. Some early examples are: *BG* 5 (1880): 13; 6 (1881): 233, 302; 8 (1883): 213; *AN* 16 (1882): 47, 315, 507–8; Dupree, *Asa Gray,* 384–95; Andrew Denny Rodgers III, *American Botany, 1873–1892: Decades of Transition* (New Brunswick, N. J., 1944), 285–320. For discussions on professionalization and science see: Edward H. Beardsley, *The Rise of the American Chemistry Profession, 1850–1900,* Univ. of Florida Monographs: Social Sciences, no. 23 (Summer 1964); Burton J. Bledstein, *The Culture of Professionalism: The Middle Class and the Development of Higher Education in America* (New York, 1976); Bruce, 3–111, 251–68, 313–56; Eugene Cittadino, "Ecology and the Professionalization of Botany in America, 1890–1905," *Studies in History of Biology* 4 (1980): 171–98; George H. Daniels, "The Process of Professionalization in American Science: The Emergent Period, 1820–1860," *Isis* 58 (1967): 151–66; Daniel J. Kevles, "American Science," in Nathan O. Hatch (ed.), *The Professions in American History* (Notre Dame, Ind., 1988), 107–25; Edward Shils, "The Order of Learning in the United States: The Ascendancy of the University," in Alexandra Oleson and John Voss (eds.), *The Organization of Knowledge in Modern America, 1860–1920* (Baltimore, 1979), 19–47; Robert H. Wiebe, *The Search for Order, 1877–1920* (New York, 1967), 111–32.

2. John M. Coulter, *Plant Relations: A First Book of Botany* (New York, 1900), 2.

3. W. F. Ganong, "Advances in Methods of Teaching: Botany," *Science* 9 (1899): 96–100; Bessey, "Science and Practice"; *AN* 14 (1880): 894; 15 (1881): 736; 16 (1882): 507; *BG* 15 (1890): 267.

4. Joseph T. Rothrock, "Home and Foreign Modes of Teaching Botany, III," *BG* 6 (1881): 233.

5. Bessey, "Modern Botany and Mr. Darwin," 507–8; *AN* 20 (1886): 254–55; *BG* 10 (1885): 216; 12 (1887): 87, 140; 13 (1888): 130; 14 (1889): 47; Garland E. Allen, *Life Science in the Twentieth Century* (New York, 1975), 1–8; P. R. Bell (ed.), *Darwin's Biological Work: Some Aspects Reconsidered* (New York, 1964), xi–xii, 1–49, 207–61; Gavin de Beer, *Charles Darwin: A Scientific Biography* (Garden City, N. Y., 1965), 225–51; Dupree, *Asa Gray,* 355.

6. J. M. Coulter to Bessey, August 24, 1880, BP.

7. Bessey, "Easily Made Observation," 172; Bessey, "The Asparagus for Histological Study," *BG* 6 (1881): 294–95; ibid., 231.
8. A. S. Packard, Jr., to Bessey, August 10, 1880, BP.
9. Bessey, "A Sketch of the Progress of Botany in the United States in the Year 1879," *AN* 14 (1880): 862–70; Bessey, "A Sketch of the Progress of Botany in the United States in 1880," ibid. 15 (1881): 947–55.
10. A. S. Packard, Jr., to Bessey, December 17, 1886, E. D. Cope to Bessey, December 26, 1886, October 3, 1887, December 16, 1891, BP; *BG* 15 (1890): 48.
11. A. S. Packard, Jr., to Bessey, October 27, 1881, March 15, 1882, BP; *BG* 6 (1881): 167, 231, 257, 277, 294–95; 7 (1882): 137; 8 (1883): 229; 9 (1884): 33.
12. *BG* 7 (1882): 51.
13. Bessey, "Observations of Frost upon Leaf-Cells," 464–65; Bessey, "*Cuscuta glomerata,*" 1145–47.
14. *BG* 8 (1883): 291–93, 296, 321; 9 (1884): 67, 84, 99, 129, 153; *AN* 17 (1883): 1065; 18 (1884): 725; 19 (1885): 802–3.
15. Joseph C. Arthur, "Botanists and Botanizing at Minneapolis," *BG* 8 (1883): 296.
16. *BG* 8 (1883): 321.
17. Ibid.
18. *BG* 9 (1884): 156–59, 174; *AN* 17 (1883): 1096.
19. *AN* 19 (1885): 170, 802–3; 24 (1890): 1082–83.
20. A. Henry to Bessey, January 9, 1885, BP; *BG* 10 (1885): 334–36; 11 (1886): 224; 13 (1888): 41; *AN* 19 (1885): 170, 604; 20 (1886): 645, 886; 21 (1887): 276–77, 376–78; 22 (1888): 46–47; 24 (1890): 675–76.
21. *AN* 19 (1885): 170.
22. Bessey, "Botanical Manuals," 376–78; Bessey, "Books on Fungi," *AN* 20 (1886): 645.
23. Bessey, "The Study of Liverworts in North America," *AN* 19 (1885): 604.
24. *AN* 22 (1888): 46–47; *BG* 13 (1888): 41.
25. Bessey, "Books on Fungi," 645; Bessey, "The Completion of Saccardo's Sylloge Fungorum," *AN* 24 (1890): 675–76; ibid., 1190.
26. *BG* 10 (1885): 336; 11 (1886): 224; 16 (1891): 348.
27. *AN* 20 (1886): 45.
28. *BG* 11 (1886): 65–66.
29. Ibid. 10 (1885): 216.
30. Ibid. 13 (1888): 130; 14 (1889): 47.
31. *AN* 20 (1886): 254–55.
32. *BG* 15 (1890): 267.
33. Ibid. 12 (1887): 87; 13 (1888): 130.
34. Ibid. 12 (1887): 140.
35. Arthur, "Botanical Laboratories," 395; *AN* 20 (1886): 281.
36. *BG* 12 (1887): 140; *AN* 20 (1886): 145.
37. *AN* 20 (1886): 537.
38. Ibid., 728.
39. Ibid.
40. Ibid. 21 (1887): 59–60.
41. H. H. Ballard to Bessey, December 16, 26, 1886, BP; *BG* 10 (1885): 338, 429; Arthur, "Botanical Laboratories," 395; Coulter, "Laboratory Appliances," 409; John M. Coulter, "Laboratory Courses of Instruction," *BG* 10 (1885): 417; ibid. 12 (1887): 140; 15 (1890): 23, 180; *AN* 19 (1885): 75; 20 (1886): 459, 536; 21 (1887): 357, 376.
42. Bessey presented a paper, "What to Do with a Beginner in Botany," at the second

meeting in 1889 and was elected president of the group at that time.
43. *AN* 22 (1888): 1043.
44. Ibid.
45. *BG* 15 (1890): 213.
46. Bessey, "Botanical Journals," *AN* 21 (1887): 79.
47. *AN* 20 (1886): 729.
48. *BG* 10 (1885): 299.
49. *AN* 23 (1889): 816.
50. Ibid., 816–19.
51. Ibid. 24 (1890): 964.
52. Ibid. 19 (1885): 802–3, 889, 997–99; 20 (1886): 888; 21 (1887): 930; 24 (1890): 958–64; *BG* 10 (1885): 336–41, 363; 11 (1886): 224–28, 279; 13 (1888): 132, 230, 242; 14 (1889): 262–64, 268; 15 (1890): 231–33; 16 (1891): 209, 261, 315.
53. *AN* 19 (1885): 889, 997; *BG* 10 (1885): 336–41, 363; 16 (1891): 261, 273.
54. *BG* 11 (1886): 248.
55. C. MacMillan to Bessey, September 22, 1892, BP; *BG* 10 (1885): 335; 11 (1886): 224; 13 (1888): 242; 14 (1889): 262, 268; 16 (1891): 261; *AN* 20 (1886): 888; 21 (1887): 930; 23 (1889): 816; 26 (1892): 833; *Science,* o.s., 19 (1892): 81, 320.
56. C. MacMillan to Bessey, September 22, 1892, BP; *BG* 11 (1886): 279; 13 (1888): 242; 14 (1889): 262, 268.
57. *BG* 16 (1891): 273; *AN* 26 (1892): 854.
58. *BG* 16 (1891): 315.
59. Ibid. 10 (1885): 299; 11 (1886): 119; 12 (1887): 197, 235; 13 (1888): 215; 14 (1889): 262; *AN* 19 (1885): 803.
60. Bessey to Professor Plum, December 26, 1891, BP; *BG* 10 (1885): 299; 13 (1888): 325; 14 (1889): 270; Byron D. Halsted, "The Society's Progress," *SPAS* (1898), 20–21, 32–42.
61. *BG* 14 (1889): 305.
62. Ibid., 306; 15 (1890): 279, 334.
63. Ibid. 16 (1891): 264; *AN* 25 (1891): 914.

## CHAPTER 5

1. *BG* 17 (1892): 285; Science, o.s., 20 (1892): 135.
2. Bessey to B. D. Halsted, January 22, August 8, 1892, Bessey to J. M. Coulter, [February 1892], Bessey to F. W. Putnam, [August 1892], Halsted to Bessey, August 25, 29, 1892, S. H. Gage to Bessey, September 28, 1892, BP.
3. Bessey, "A Plea for the Better Pronunciation of Botanical Names," Botanical Club, AAAS; Bessey to B. Wheeler, June 21, 1893, A. Wright to Bessey, October 15, 1894, Bessey to Wright, October 18, 1894, Bessey to Members of the Club, August 14, 1895, BP.
4. Bessey to N. Britton and B. Robinson, January 6, 1897, BP; *AN* 31 (1897): 151–52.
5. F. V. Coville to J. S. Morton, November 23, 1893, Bessey to J. M. Thurston, April 3, 1897, BP.
6. L. Underwood to S. Watson, December 12, 1888, October 16, 30, 1889, Underwood to G. Davenport, September 9, 1901, Underwood folder, D. Eaton to G. Davenport, November 18, 1892, Eaton folder, Gray Herbarium, Archives, Harvard University, Cambridge, Mass.; J. Redfield to N. Britton, October 5, 1888, Redfield folder, D. Eaton to E. Britton, January 3, 1889, Eaton folder, Nathaniel Britton Correspondence, Library of the

New York Botanical Garden, Bronx, N.Y.; Dupree, *Asa Gray,* 387–402.

7. M. Bebb to S. Watson, July 21, 1889, February 4, 1890, December 16, 1894, Bebb folder, Watson to L. Underwood, October 19, 1889, Underwood folder, J. M. Coulter to A. Gray, December 3, 1887, Coulter to Watson, April 18, 27, May 16, 1888, October 26, November 26, December 1, 1891, Coulter folder, N. Britton to Watson, April 18, 1888, April 29, 1890, Britton folder, D. Eaton to Watson, February 3, 11, March 30, 1890, Eaton folder, Gray Herbarium; Watson to N. Britton, April 1, 11, 1888, June 12, 1890, Watson folder, Britton Corresp.

8. N. Britton to A. Gray, November 30, 1887, Britton folder, Gray Herbarium; S. Watson to N. Britton, December 6, 1887, Gray folder, Britton Corresp.; *Science,* o.s., 12 (1888): 101.

9. L. Underwood to G. Davenport, March 14, 1892, Underwood folder, Gray Herbarium; Underwood to N. Britton, April 26, 1892, Underwood folder, Materials on Rochester Code, Letters and Documents on Nomenclature, 1892–98, Misc. Box, Britton Corresp. (cited as Nomenclature, Britton).

10. Bessey to B. D. Halsted, August 8, 1892, BP.

11. There are numerous items pertaining to the work of the checklist committee in Nomenclature, Britton; *BG* 17 (1892): 285–87, 297; *AN* 26 (1892): 854.

12. *BG* 17 (1892): 287–88, 297.

13. J. C. Arthur to B. T. Galloway, June 1, 1892, Erwin F. Smith Papers, American Philosophical Society Library, Philadelphia; *BG* 17 (1892): 22, 287–89, 297, 339.

14. L. Underwood to G. Davenport, October 23, 1892, Underwood folder, Gray Herbarium.

15. *BG* 17 (1892): 297; *AN* 26 (1892): 854; *Science,* o.s., 19 (1892): 146–47.

16. L. Underwood to N. Britton, October 13, 1892, Nomenclature, Britton.

17. *BG* 17 (1892): 339, 341.

18. Bessey to W. Wilson, December 9, 1892, Bessey to J. C. Arthur, December 9, 1892, BP; Arthur to S. Watson, March 10, 1890, Coulter folder, Gray Herbarium; Arthur to N. Britton, May 12, 1890, Arthur folder, Britton Corresp.; *BG* 17 (1892): 304, 425; 18 (1893): 36; AN, 27 (1893): 297.

19. *BG* 17 (1892): 297, 383; 18 (1893): 316; *Science,* o.s., 19 (1892): 320.

20. Bessey to O. Kuntze, July 7, 31, 1893, BP; *BG* 18 (1893): 316; *AN* 27 (1893): 48–49; *Science,* o.s., 20 (1892): 164, 219, 245; 21 (1893): 105.

21. *BG* 17 (1892): 297; 19 (1894): 126; AN, 26 (1892): 147–55, 226–31; 28 (1894): 1030.

22. *Science* 3 (1896): 13–16; *BG* 18 (1893): 342; *AN* 27 (1893): 821.

23. *AN* 27 (1893): 823.

24. Bessey to C. Barnes, December 10, 1892, Bessey to N. Britton et. al., February 1893, BP.

25. *BG* 18 (1893): 333; *AN* 27 (1893): 823–28.

26. *AN* 27 (1893): 823.

27. Ibid., 823–28, 913–16; *BG* 18 (1893): 342, 368. There are numerous items in 1893 regarding the report of the committee in Nomenclature, Britton Corresp.

28. *BG* 18 (1893): 194, 316; 19 (1894): 126; *AN* 28 (1894): 1030.

29. Botanical Society of America folder, William Trelease Papers, Missouri Botanical Garden Archives and Manuscript Collections, St. Louis; C. Barnes to B. Robinson, May 18, 1895, Barnes folder, L. Underwood to S. Watson, October 16, 1889, Underwood to G. Davenport, September 9, 1901, Underwood folder, Gray Herbarium; J. Redfield to N. Britton, October 5, 1888, June 11, 1890, Redfield folder, Britton Corresp.; *BG* 18 (1893): 349; *AN* 27 (1893): 823–28. Of the eight whose views constituted the majority report that

opposed establishing the society at that time, only Lester Frank Ward refused to join and to give his support to the organization once it was founded (L. F. Ward to W. Trelease, November 27, 1893, Trelease Papers).

30. *BG* 18 (1893): 342, 368; *AN* 27 (1893): 913–16.
31. *BG* 18 (1893): 350–55; *AN* 27 (1893): 826–28.
32. Ibid.
33. Bessey to G. Goodale, August 29, 1893, BP; *AN* 27 (1893): 828.
34. *BG* 17 (1893): 364.
35. Ibid. 18 (1894): 428.
36. *Science* 1 (1895): 139; 8 (1898): 651; 12 (1900): 577.
37. *BG* 19 (1894): 126; 20 (1895): 97, 162, 180; *AN* 28 (1894): 292; 29 (1895): 349–51.
38. C. Barnes to B. Robinson, May 16, 1895, Barnes folder, Gray Herbarium; *BG* 20 (1895): 426.
39. *BG* 20 (1895): 426–27; *AN* 29 (1895): 373–75.
40. A. Chapman to B. Robinson, March 27, 1895, Chapman folder, Robinson to G. Davenport, April 9, 1895, Robinson folder, Gray Herbarium; F. Coville to N. Britton, September 3, 1895, Nomenclature, Britton; W. Farlow to W. Trelease, April 22, 1895, William Trelease Papers, Department of Manuscripts and University Archives, Cornell University Library, Ithaca, N. Y.; *BG* 20 (1895): 263; 21 (1896): 85.
41. N. Britton to Committee on Nomenclature, August 28, 1895, Nomenclature, Britton; Britton to Committee on Nomenclature, September 18, 1895, Box 6, Liberty Hyde Bailey Papers, Department of Manuscripts and University Archives, Cornell University Library; *BG* 20 (1895): 414; *Science* 2 (1895): 445.
42. F. Coville to Bessey, June 10, 1895, BP; *BG* 20 (1895): 428. A detailed statement, given privately, in opposition to the Harvard Circular was: C. Barnes to B. Robinson, May 18, 1895, Barnes folder, Gray Herbarium. A number of items regarding this are in 1895, Nomenclature, Britton.
43. F. Coville to Bessey, June 10, August 20, 1895, BP.
44. Bessey to A. J. McClatchie, [August 1892], BP.
45. Bessey to N. Britton, October 26, 1894, BP.
46. *AN* 29 (1895): 349–51.
47. N. Britton to Bessey, October 22, 1896, BP.
48. Bessey to G. P. Clinton, September 24, 1894, Bessey to N. Britton, August 16, 1895, BP; Bessey, "A Protest Against the 'Rochester Rules,' " *AN* 29 (1895): 666–68.
49. Ibid.
50. 50. Bessey to B. Halsted, July 18, 1892, Bessey to A. Engler, August 8, 1892, N. A. Hawey to Bessey, December 4, 1892, Bessey to Hawey, December 6, 1892, J. B. Ham to Bessey, January 26, 1893, Bessey to C. MacMillan, August 26, 1895, BP; *AN* 26 (1892): 1022.
51. Bessey to N. Britton, August 14, 1895, BP.
52. A. Woods to Bessey, [1894], BP.
53. R. K. Beattie to Bessey, October 24, 1898, BP.
54. M. E. Jones to Bessey, May 24, 1896, BP.
55. L. Underwood to G. Davenport, January 13, December 15, 1895, Underwood folder, Gray Herbarium.
56. *AN* 27 (1893): 365–66; *BG* 21 (1896): 177, 296; 24 (1897): 120; Nathaniel Lord Britton and Addison Brown, *An Illustrated Flora of the Northern United States, Canada, and the British Possessions* (New York, 1896).
57. *BG* 22 (1896): 338.
58. Botanical Society of America folder, Trelease Papers, Missouri Botanical Garden; *BG* 18 (1893): 349; 21 (1896): 338; *AN* 27 (1893): 823–28; Oswald Tippo, "The Early

History of the Botanical Society of America," in Steere, 1-13.
59. D. Eaton to W. Trelease, October 15, 1893, Trelease Papers, Missouri Botanical Garden.
60. Ibid., March 12, 1894.
61. Ibid.
62. Ibid.; R. Thaxter to W. Trelease, [n.d.] and January 1, 1894, Trelease Papers, Missouri Botanical Garden.
63. W. Farlow to W. Trelease, June 12, October 5, November 21, 1893, October 12, November 14, 1894, Trelease Papers, Cornell University.
64. B. Robinson to W. Trelease, December 20, 1893, ibid.
65. R. Thaxter to W. Trelease, [n.d.] and January 1, 1894, C. Sargent to Trelease, September 4, 1893, Trelease Papers, Missouri Botanical Garden.
66. Botanical Society of America folder, ibid.
67. C. Barnes to members, 1895, ibid.; Barnes to Bessey, July 21, December 23, 1895, BP.
68. G. Atkinson to members, July 31, 1900, Bessey to D. T. MacDougal, November 13, 1903, BP; *BG* 24 (1899): 210.
69. C. Barnes to Bessey, February 22, 1898, BP; *BG* 20 (1895): 403; 22 (1896): 218; 24 (1897): 179; 26 (1898): 296; 28 (1899): 210.
70. L. H. Bailey to C. Barnes, January 22, 1895, W. Trelease to C. Sargent, February 12, 1895, Barnes to Trelease, January 26, 1895, Trelease Papers, Missouri Botanical Garden.
71. C. Barnes to W. Trelease, April 2, 1895, ibid.
72. Bessey to C. Barnes, June 27, 1895, BP.
73. Bessey to G. Atkinson, March 2, 19, May 9, 1898, BP.
74. Bessey to W. Trelease, March 25, 1898, Bessey to G. Atkinson, March 26, May 9, 1898, BP.
75. C. Barnes to W. Trelease, February 2, 1897, Barnes to members, February 15, 1897, Trelease Papers, Missouri Botanical Garden; B. Halsted to members, July 27, 1900, Bessey to Halsted, August 3, 1900, BP.
76. G. Atkinson to members, July 20, 1900, BP; *BG* 28 (1899): 210.
77. Toby A. Appel, "Organizing Biology: The American Society of Naturalists and Its 'Affiliated Societies,' 1883-1923," in Ronald Rainger, Keith R. Benson, and Jane Maienschein (eds.), *The American Development of Biology* (Philadelphia, 1988), 87-120.
78. Committee statement to botanists, June 1, 1896, Trelease Papers, Missouri Botanical Garden; J. E. Humphrey to Bessey, [1895] and January 19, 1896, Bessey to Humphrey, October 31, 1896, Bessey to C. Barnes, November 16, 1896, BP; Appel, 102.
79. Bessey, "Relative Infrequence of Fungi upon the Trans-Missouri Plains and the Adjacent Foothills of the Rocky Mountain Region," *AN* 33 (1899): 215.
80. E. D. Cope to Bessey, April 5, 1895, BP.
81. E. D. Cope to Bessey, March 17, April 5, 1895, BP; Cope to E. F. Smith, April 25, 1895, Smith Papers.
82. Bessey to N. Britton, August 14, 1895, BP; C. R. Barnes to E. F. Smith, June 11, 1895, N. L. Britton to E. F. Smith, August 12, 1895, Smith Papers; *Science* 2 (1895): 412.
83. *BG* 20 (1895): 334.
84. Bessey to unknown, July 21, 1897, Bessey to J. M. Cattell, July 21, September 25, 1897, Bessey to W. G. Farlow, October 18, 1897, Cattell to Bessey, August 29, September 16, 1897, BP.
85. Bessey to J. M. Cattell, March 1, 1901, January 7, 1904, Cattell to Bessey, March 14, 1901, January 26, 1904, BP.
86. Bessey to H. Bolley, February 12, 1896, A. F. Woods to Bessey, November 24,

1902, BP; *SPAS* (1881), 9–11; I. P. Roberts, "The Promotion of Agricultural Science," ibid. (1897), 82–85; W. H. Jordan, "The Promotion of Agricultural Science," ibid. (1902), 22–33; *AN* 20 (1886): 65–66; Shils, 23–24; Margaret W. Rossiter, "The Organization of the Agricultural Sciences," in Oleson and Voss, 211–48.

87. Committee of M. A. Scovell to members, November 26, 1894, Executive Committee of SPAS to members, March 17, 1896, C. S. Plumb to members, September 28, 1897, Bessey to Plumb, May 26, 1899, BP; *SPAS* (1899), 18.

88. H. L. Bolley to Bessey, January 14, 1896, BP; *Science* 8 (1898): 761.

89. *SPAS* (1897), 14.

90. Bessey to C. S. Plumb, May 26, 1899, BP; *Science* 9 (1899): 689.

91. J. C. Arthur to Bessey, May 9, 1907, BP; *AN* 24 (1890): 94; *SPAS* (1905), 1–16; (1906), 2–5; (1907), 2–5; (1908), 20; (1909), 23–25; (1911), 5–9; (1912), 5–8; (1916), 6; (1917), 5; (1919), 8; Charles E. Thorne, "Our Place in the Sun," ibid. (1916), 6–14; E. W. Allen, "The Society for the Promotion of Agricultural Science: Its Present and Its Future," ibid. (1920), 96–99; *Science* 57 (1923): 216.

92. George H. Daniels, "The Pure-Science Ideal and Democratic Culture," *Science* 156 (1967): 1699–1701; I. Bernard Cohen, "Science in America: The Nineteenth Century," in Arthur M. Schlesinger, Jr. and Morton White (eds.), *Paths of American Thought* (Boston, 1963), 181–88.

93. Bessey, "Science and Practice;" Bessey, "Some of the Next Steps in Botanical Science," *Science* 37 (1913): 5.

94. Nathan Reingold, "Alexander Dallas Bache: Science and Technology in the American Idiom," *Technology and Culture* 11 (April 1970): 163–77; Edwin Layton, "Mirror-Image Twins: The Communities of Science and Technology in 19th-Century America," ibid. 12 (October 1971): 562–80.

95. Jordan, 26.

96. Bessey, "Some of the Next Steps," 3, 7–11; Bessey, "Plant Physiology in Agricultural Education," 50–54.

97. Bessey, "Some of the Next Steps." 9–10.

98. Ibid., 4–5.

99. Bessey, "Laying the Foundations," 40–44.

100. R. Pound to N. Britton, September 13, 1897, Nomenclature, Britton; *BG* 30 (1900): 403.

101. Various items and letters to and from Britton, 1900–1906, Nomenclature, Britton; *BG* 34 (1902): 157.

102. F. Coville to N. Britton, July 19, August 24, 1905, Britton to F. Earle, August 15, 1905, Nomenclature, Britton; *Science* 22 (1905): 217–19, 468–69; *BG* 40 (1905): 68–72; 42 (1906): 493.

103. *BG* 40 (1905): 72.

104. Minutes of Committee of Nomenclature, June 30, 1902, Report of Committee of Nomenclature to Botanical Club 1902 (January 1903), C. Pollard to N. Britton, January 22, 1903, Proposals of Systematic Seminar of Washington, D.C., to Nomenclature Committee (1903), Report of Nomenclature Commission to Botanical Club, December 26, 1903, "A Proposed Code of Botanical Nomenclature," F. Coville to Britton, January 19, 21, February 8, 1904, Britton to A. Hollick, January 27, February 15, 1904, Code of Botanical Nomenclature, February 6, 1904, Britton to J. Briquet, May 9, 1904, Nomenclature, Britton; *BG* 44 (1907): 80, 240; 46 (1908): 467–68; 47 (1909): 153; 49 (1910): 149–50; 50 (1910): 220–22.

105. Examples of this in ecology and plant breeding are Cittadino, 171–98; and Diane

B. Paul and Barbara A. Kimmelman, "Mendel in America: Theory and Practice, 1900–1919," in Rainger, Benson, and Maienschein, 281–310.

106. *BG* 33 (1902): 87; 35 (1903): 152.

107. Bessey to L. M. Underwood, December 11, 1903, Bessey to C. L. Barnes, December 17, 1903, Bessey to G. Atkinson, February 2, 1904, BP.

108. Beverly T. Galloway, "The Twentieth Century in Botany," *Science* 19 (1904): 11–18.

109. B. T. Galloway to Bessey, April 4, 1904, Galloway to C. Barnes and Bessey, September 15, 1904, Bessey to Galloway, July 15, December 23, 1904, Bessey to G. Atkinson, October 24, 1904, Bessey to Barnes, December 12, 1905, W. Trelease to Bessey, December 20, 1905, BP; N. Britton to B. Robinson et. al., October 26, November 18, 23, 1905, Britton folder, Gray Herbarium; numerous letters between Atkinson and Britton, C. Shear, W. Ganong, Bessey, F. Clements, D. MacDougal, copies of proposed constitutions, September 1904–January 1906, George Atkinson Papers, Department of Manuscripts and University Archives, Cornell University Library, Ithaca, N.Y.

110. Bessey to C. Barnes, December 17, 1903, W. Trelease to Bessey, December 20, 1905, BP.

111. For more on the continuing elusiveness of professional homes for the various fields of biology see Appel.

112. W. Ganong to C. Barnes, November 17, 1903, enclosed in letter, B. T. Galloway to Bessey, April 4, 1904, BP.

## CHAPTER 6

1. Bessey to S. Aughey, November 25, 1892, BP.

2. Herbert J. Webber, "The Flora of Central Nebraska," *AN* 23 (1889): 633.

3. B. D. Halsted to Bessey, June 14, July 18, 1887, December 10, 1889, BP. There are numerous examples in the Bessey Papers of field trips in conjunction with speeches; such as C. C. Williamson to Bessey, May 17, 1893, BP.

4. *Daily Nebraska State Journal,* December 14, 1888; H. K. W., "The Botanical Seminar," *University Journal,* 3 (December 1906): 34; Bessey to G. Vasey, March 14, 1892, Bessey to P. A. Rydberg, April 19, 1892, April 22, 1893, Bessey to R. Furnas, June 20, 1892, Bessey to J. S. Morton, March 18, 1893, Bessey to F. Coville, July 3, 1893, Bessey to F. C. Newcombe, June 14, 1894, Bessey to J. A. Lendbo, February 6, 1897, "The Book of the Sem. Bot.," n.d., BP; Sem Bot Club Records, Misc. Items (1), University of Nebraska–Lincoln Archives; Ronald C. Tobey, *Saving the Prairies: The Life Cycle of the Founding School of American Plant Ecology, 1895–1955* (Berkeley, 1981), 14–21.

5. Thomas A. Williams, "Notes on Nebraska Lichens," *AN* 23 (1889): 161; Herbert J. Webber, "The Fresh-Water Algae of the Plains," ibid., 1011–13; Webber, "Central Nebraska," 633–36; Williams, "Notes on the Canyon Flora of Northwest Nebraska," ibid. 24 (1890): 779; Per Axel Rydberg, "The Flora of the High Nebraska Plains," ibid., 25 (1891): 485; Williams, "Notes on the Flora of Western South Dakota," ibid., 26 (1892): 60, 253.

6. Bessey to H. Webber, September 9, 1891, Bessey to J. M. Bates, October 14, 1891, Bessey to I. Newford, June 1, 1892, BP; Bessey and Webber, "Grasses and Forage Plants and Catalogue," 144–74; Webber, "Appendix to the Catalogue of the Flora of Nebraska," *Neb Bd Agr* (1892), 45–53; Bessey, "Supplementary List of Recently Reported Species," in Webber, ibid.; Webber, "Central Nebraska," 633; *BG* 17 (1892): 132, 228. Also at this time G. D. Swezey added to the knowledge of the state by publishing a listing of the flowering

plants as represented in his herbarium at Doane College (*BG* 16 [1891]: 188; Swezey, "Additions to the Flora of Nebraska," *Publications of the Nebraska Academy of Sciences* 2 [1892]: 16–17 [cited as *Neb Acad Sci*]).

7. Bessey, "The Eastward Extension of *Pinus ponderosa* Douglas, var. scopulorum," *AN* 21 (1887): 928–29; Bessey, "The Westward Extension of the Black Walnut," ibid., 929; Bessey, "The Iron-Weed Tree in the Black Hills," ibid.; Bessey, "Grass Flora of Nebraska Plains," 171–72; Bessey, "The Bearberry in Central Nebraska," ibid. 25 (1891): 1130.

8. Bessey, "A Meeting-Place for Two Floras," *Bulletin of the Torrey Club* 4 (1887): 189–91; Bessey, "The Flora of the Upper Niobrara," *AN* 23 (1889): 537–38; Bessey, "Notes on the Flora of the Black Hills," *Neb Acad Sci* 2 (1892): 17–19; Bessey, "Extension of *Pinus ponderosa*," 928–29.

9. H. J. Webber to Bessey, September 5, 1892, Bessey to F. Coville, January 16, 1894, Bessey to F. Scribner, May 23, 1894, Scribner to Bessey, March 13, 1895, A. Woods to Bessey, December 15, 1896, note, December 8, [n.d.], folder 1, General Botany Notes, BP.

10. Bessey, "Progress of the Botanical Survey of Nebraska," *AN* 29 (1895): 580; Bessey, "Grasses and Other Forage Plants," 79–93; Roscoe Pound, "Progress of the Botanical Survey of Nebraska," *Neb Acad Sci* 4 (1894): 7–8, 138; J. E. Weaver and Albert F. Thiel, "Ecological Studies in the Tension Zone between Prairie and Woodland," *Botanical Survey of Nebraska,* n.s., no. 1 (April 1917), 60.

11. Note [January 1894], P. A. Rydberg to F. Coville, February 28, 1894, Division of Botany.

12. P. A. Rydberg to F. Coville, January 17, 1894, ibid.; Bessey to G. Vasey, February 23, 1893, Coville to Bessey, April 17, 19, May 27, 1893, Bessey to Coville, April 25, 1893, Rydberg to Bessey, June 30, 1893, R. Pound to Bessey, July 10, 24, 1893, BP.

13. F. L. Scribner to Bessey, March 11, 1896, BP.

14. *BG* 18 (1893): 78; Pound, 7.

15. *BG* 19 (1894): 469; *Science* 1 (1895): 25.

16. G. E. MacLean to Bessey, December 17, 1898, BP; *BG* 21 (1896): 177; 28 (1899): 79.

17. Ibid. 21 (1896): 177.

18. C. L. Shear to F. Coville, December 5, 1894, P. A. Rydberg to Coville, May 13, 1895, F. Clements to Coville, April 24, 1897, Division of Botany; *Science* 4 (1896): 964; *AN* 30 (1896): 61; 34 (1900): 77; *BG* 29 (1900): 360.

19. Bessey to Porter, September 6, 1894, Bessey to J. Macoun, February 21, 1898, Bessey to A. S. Hitchcock, April 12, 1898, BP.

20. *BG* 18 (1893): 333; *AN* 31 (1897): 652–54.

21. Bessey, "Notes on the Distribution of the Yellow Pine in Nebraska," *Garden and Forest* 8 (1895): 102–3; Bessey, "Are the Trees Receding from the Nebraska Plains?" ibid. 10 (1897): 456–57; Bessey, "The Origin of the Flora of Nebraska," *Neb Acad Sci* 5 (1896): 11, 33.

22. All by Bessey, "Preliminary Description of Native and Introduced Grasses," 209–79; "Grasses and Other Forage Plants," 79–93; "Forage Problem," 107–20; "Grasses of Nebraska," 175–88; "The Weeds of Nebraska," *SPAS* (1893), 33–44; "The Russian Thistle in Nebraska," ibid. (1894), 286; "Buffalo Grass," 105–6; "Passing of Russian Thistle," 83–85.

23. Roscoe Pound, "The Plant-Geography of Germany," *AN* 30 (1896): quoted 465, 465–68.

24. Ibid., 465.

25. Ibid., 466; Tobey, *Saving the Prairies,* 48–75.

26. Roscoe Pound and Frederic E. Clements, *The Phytogeography of Nebraska, A*

*General Survey* (Lincoln, 1898); Pound and Clements, "The Vegetation Regions of the Prairie Province," *BG* 25 (1898): 381–94.

27. Bessey, "Some Characteristics of the Foothill Vegetation of Western Nebraska," *AN* 32 (1898): 111–13.

28. *Report of the Commissioner of Agriculture,* USDA (1886), 152, 165–66, (1888), 603–18; *Report of the Secretary of Agriculture,* USDA (1889), 276; J. K. Macomber, "The Relation of Forest to Climate," *Ia Hort Soc* 10 (1875): 241–47; J. E. Todd, "The Relation of Forests to Rainfall," ibid. 13 (1878): 112–18; Macomber, "Adaptability of Prairie Soils for Timber Growth," ibid. 14 (1879): 292–97; J. D. Whitney, "Plain, Prairie, and Forest," *AN* 10 (1876): 577; O. P. Hay, "An Examination of Prof. Leo Lesquereux's Theory of the Origin and Formation of Prairies," ibid. 12 (1878): 299; J. E. Todd, "Notes on the Distribution of Timber in South-Western Iowa, with Inferences Concerning the Origin of Prairies," ibid., 91; Robert W. Furnas, "Tree Growth on the Plains," ibid. 20 (1886): 380–81; C. A. White, "Adaptability of the Prairies for Artificial Forestry," *Science,* o.s., 3 (1884): 438–42; Charles R. Kutzleb, "Can Forests Bring Rain to the Plains?" *Forest History* 15 (1971): 14–21; David M. Emmons, "Theories of Increased Rainfall and the Timber Culture Act of 1873," ibid., 6–14.

29. Macomber, "Relation of Forest to Climate," 242.

30. R. N. Kratz to Bessey, February 25, 1887, BP; H. H. McAfee, "Report of Committee on Forestry," *Ia Hort Soc* 9 (1874): 135–47.

31. W. H. Brewer to Bessey, January 19, 1876, J. D. Dana to Bessey, May 27, 1876, R. W. Furnas to Bessey, April 15, 1886, Bessey to Professor Keffer, [September 1894], BP.

32. *Botanical Survey of Nebraska,* 1–7 (Lincoln, 1892–1904); Bessey, "Report Upon the Native Trees and Shrubs of Nebraska," *Neb Exp St,* Bulletin #18 (1891); Bessey, "Progress of Botanical Survey," 580.

33. Bessey, "The Condition of Forests and Forestry: Nebraska," *Proceedings of the American Forestry Association* 2 (1896): 83–89; Bessey, "Second Report upon Native Trees," 154–85; Bessey, "Were the Sand Hills of Nebraska Formerly Covered with Forests?" *Neb Acad Sci* 5 (1894): 7; Bessey, "The Box-Elder on the Plains," *Garden and Forest* 9 (1896): 33.

34. Ibid.; Bessey to A. F. Woods, December 15, 1894, J. Adamson to Bessey, June 10, 1895, Bessey to E. L. Mark, January 4, 1898, BP; J. S. Kingsley, "The Hat Creek (Nebraska) Bad Lands," *AN* 25 (1891): 963–64; E. M. Hussong, "The Yellow Pine in the Republican Valley," *Neb Acad Sci* 5 (1894): 7–8; P. A. Rydberg, "Flora of the Sand Hills of Nebraska," *Contributions from the United States National Herbarium,* no. 3, 3 (1895): 145–46; *BG* 20 (1895): 552; White, 438–43.

35. B. E. Fernow to Bessey, November 3, 1890, February 12, 1891, BP; Bessey, "Report of the Botanist," *Neb Bd Agr* (1890), 37–40; *Report of the Chief of the Division of Forestry,* USDA (1891), 206; (1892), 298–301; Raymond J. Pool, "Fifty Years on the Nebraska National Forest," *Nebraska History* 34 (1953): 144–45.

36. Bessey to J. C. Toliver, November 16, December 9, 1893, BP.

37. Quoted in Pool, "Fifty Years," 145.

38. Ibid.

39. Bessey to W. G. Hawkins, November 12, 1892, O. A. Davis to Bessey, [1892], Bessey to J. C. Toliver, November 16, 1893, Bessey to G. Heimer, November 12, December 21, 1894, R. A. Emerson to Bessey, December 3, 1900, folder 2, Botany General Notes, BP.

40. J. C. Toliver to Chancellor Canfield, November 3, 1893, Toliver to Bessey, December 12, 26, 1893, March 6, 1894, BP.

41. J. C. Toliver to Bessey, May 4, July 25, 1894, Bessey to Toliver, September 10, 1897, BP.

42. Bessey to J. C. Toliver, April 6, 1894, BP.
43. Bessey to B. E. Fernow, December 24, 1894, January 10, 22, 1895, Corresp. of the Forest Service, RG 95, National Archives; Kingsley, 963–64.
44. Bessey to B. E. Fernow, December 3, 1901, BP.
45. Bessey, "Distribution of Yellow Pine," 102–03.
46. Bessey, "Condition of Forests," 84; J. Pedersen-Bjergaard, "Forest Planting on the Sands of Denmark," *Amer Agr* 52 (1893): 19.
47. Bessey, "Are Trees Receding?" 456–57; Bessey, "Distribution of Forest Trees on the Nebraska Plains," *Atlantic Slope Naturalist* 1 (1903): 21; Bessey, "Condition of Forests," 83–89; Bessey, "Grasses and Forage Plants," 204; Bessey, "Twenty Native Forest Trees of Nebraska," *Forester* 7 (1901): 314–17; *BG* 18 (1893): 333; *AN* 31 (1897): 652–54.
48. Bessey, "Second Report upon Native Trees," 184.
49. C. M. Van Metre to Bessey, March 12, 1898, C. H. Barnard to Bessey, July 3, 1899, G. S. Christy to Bessey, [1899], BP; Bessey, "Distribution of Yellow Pine," 102–3; Bessey, "Are Trees Receding?" 456–57; Bessey, "Trees on the Nebraska Plains—Are They Increasing or Diminishing from Natural Causes?" *Neb Bd Agr* (1897), 313–16; Bessey, "The Natural Spreading of Timber Areas," *Forester* 6 (1900): 240–43.
50. Bessey, "Natural Spreading," 240–41.
51. J. W. Blankenship to Bessey, June 25, August 1, 1900, October 17, 1901, Bessey to Blankenship, October 19, 1901, BP.
52. Bessey, "Natural Spreading," 241.
53. Ibid., 241–42.
54. Ibid., 242–43.
55. Bessey, "Distribution of Forest Trees," 21.
56. Bessey, "Plant Migration Studies: Forest Trees," *University Studies of the University of Nebraska* 5 (1905): 11–37.
57. Cittadino, 171–98; H. A. Gleason, "Twenty-Five Years of Ecology, 1910–1935," *Brooklyn Botanic Garden Memoirs* 4 (1936): 41–49; Andrew Denny Rodgers III, *Bernhard Eduard Fernow: A Story of North American Forestry* (New Brunswick, N. J., 1951), 84, 367; Paul B. Sears, "Plant Ecology," in Joseph Ewan (ed.), *A Short History of Botany in the United States* (New York, 1969), 124–25; Sears, "Botanists and Conservation," 359–66; V. M. Spalding, "The Rise and Progress of Ecology," *Science* 17 (1903): 201–10; Ronald Tobey, "Theoretical Science and Technology in American Ecology," *Technology and Culture* 17 (1976): 718–28; Tobey, *Saving the Prairies,* 9–47.
58. Bessey, "Migration Studies," 11–12, 28–30; Bessey, "Natural Spreading," 240–43; Tobey, "American Ecology," 720–24.
59. A. F. Woods to Bessey, October 5, 1905, BP.
60. H. J. Webber to Bessey, September 5, 1892, November 12, 1894, P. J. O'Gara to Bessey, December 4, 1902, January 17, 1903, Bessey to H. L. Shantz, January 10, 1903, Bessey to A. F. Woods, June 18, 1903, B. T. Galloway to Bessey, December 8, 1903, Bessey to Galloway and others, December 11, 1903, H. Metcalf to Bessey, May 9, 1904, Bessey to J. L. Sheldon, May 28, 1904, Bessey to H. C. Cowles, December 16, 1908, E. Field to Bessey, December 5, 1909, BP.
61. Bessey to A. F. Woods, June 24, 1903, BP.
62. Bessey to F. E. Lloyd, February 9, 1906, BP.
63. Tobey, *Saving the Prairies,* 48–109.
64. F. E. Clements to Bessey, July 28, 1901, A. G. Tansley to Bessey, December 15, 1905, BP; *Science* 13 (1901): 472; 16 (1902): 160; Rodgers, *Fernow,* 413.
65. F. E. Clements to Bessey, December 14, 1914, BP.
66. J. E. Tilden to Bessey, February 9, 1907, BP.

67. *Science* 8 (1898): 372–73; 10 (1899): 857–58.
68. For a general consideration of Bessey and conservation, see Thomas R. Walsh, "The American Green of Charles Bessey," *Nebraska History* 53 (1972): 35–57.
69. Bessey to W. J. Bryan, December 18, 1894, BP.
70. Bessey to C. S. Sargent, October 16, 1897, BP.
71. Bessey to B. E. Fernow, February 22, 1892, February 6, 1893, December 24, 1894, January 10, 22, 1895, Corresp. of Forest Service; J. A. Warder to Bessey, February 28, 1882, Fernow to Bessey, September 20, 1887, November 3, 1890, February 12, 1891, J. C. Toliver to Bessey, December 26, 1893, C. Callahan to Bessey, November 25, 1895, Bessey to W. L. Hall, December 5, 1904, BP; *Report of the Secretary of Agriculture,* USDA (1891), 195; *Neb Bd Agr* (1897), 336–52; *Forester* 5 (1899): 237; Sears, "Botanists and Conservation," 360.
72. Bessey, "Distribution of Yellow Pine"; Bessey, "Condition of Forests"; Bessey, "Box-Elder"; Bessey, "Are Trees Receding?"
73. Bessey, "Sargent's Studies on the Forests of Japan," *AN* 29 (1895): 1049–56; Bessey, "Distribution of Native Forest Trees of Nebraska," *Neb Bd Agr* (1903), 257–80; Bessey, "Trees Increasing or Diminishing," 313–15.
74. G. Pinchot to Bessey, August 26, 1899, June 28, July 19, October 25, December 9, 31, 1901, J. W. Toumey to Bessey, February 3, 1900, BP.
75. Bessey to F. H. Newell, December 14, 1897, Newell to Bessey, February 19, 1898, Bessey to W. O. Jones, February 13, 1899, Bessey to C. S. Harrison, February 16, 1899, E. A. Bowers to Bessey, December 18, 1903, BP; *Nebraska Farmer* (1899), 114; *Forester* 6 (1900): 40.
76. Bessey to W. V. Allen, April 8, 1898, Bessey to F. M. Pollard, January 28, 1909, BP; *Neb Hort Soc* (1887), 37–38.
77. Bessey to J. C. Toliver, September 10, 1897, BP.
78. Copy of letter from W. L. Hall, December 18, 1900 in T. C. Jackson to Bessey, December 24, 1900, BP; *Report of the Secretary of Agriculture,* USDA (1891), 206–11; (1892), 298–301; Pool, "Fifty Years," 142–49; Walsh, "American Green," 36–44.
79. W. L. Hall to Bessey, March 29, 1901, Report of the Department of Botany, April 1901, Bessey to Hall, April 4, 13, December 6, 1901, Bessey to R. S. Kellogg, August 9, October 8, 1901, Kellogg to Bessey, August 12, 1901, BP; *Forester* 4 (1898): 226; 7 (1901): 98–99; Pool, "Fifty Years," 146–48.
80. C. M. Van Metre to Bessey, March 19, 1901, Bessey to Van Metre, March 27, 1901, C. H. Barnard to Bessey, May 9, 1901, W. L. Hall to Bessey, July 10, 1901, January 13, 1902, Bessey to Governor Savage, September 25, 1901, Bessey to Hall, September 30, 1901, E. A. Boehne to Bessey, January 30, 1902, BP; *Forester* 7 (1901): 207–8.
81. "Report of the Bureau of Forestry on Investigations in Nebraska," January 10, 1902, attached to W. L. Hall to Bessey, January 13, 1902, BP; William L. Hall, "The Investigation Now Being Made in Nebraska by the U. S. Bureau of Forestry," *Forester* 7 (1901): 188–93; Gifford Pinchot, "The Immediate Future in Forest Work," *Forestry and Irrigation* 8 (1902): 18–21.
82. Quotations from Pool, "Fifty Years," 144–46.
83. Bessey to J. C. Toliver, November 29, 1901, Bessey to W. L. Hall, December 6, 1901, BP.
84. W. L. Hall to Bessey, January 13, 1902, BP; *Forestry and Irrigation* 8 (1902): 50.
85. Bessey to G. Pinchot, January 25, 27, 1902, Pinchot to Bessey, January 28, 1902, Bessey to B. Irwin, February 12, 1902, BP.
86. Ibid.; Bessey to President of the U. S., January 25, 1902, BP.
87. *Report of the Bureau of Forestry,* USDA (1905), 203; Pool, "Fifty Years," 149.

88. Bessey to Senator Dietrich, April 11, 1902, Bessey to G. Pinchot, January 17, 1903, BP; *Forestry and Irrigation* 8 (1902): 393; 9 (1903): 519–20.
89. Bessey to G. Pinchot, January 17, 1903, Bessey to W. L. Hall, January 20, 1903, BP.
90. C. A. Scott to Bessey, April 25, 1903, BP.
91. J. Rebmann to Bessey, February 23, March 16, September 20, 1902, February 9, 1903, BP.
92. C. W. Edgerton to Bessey, May 16, 1904, BP.
93. C. Bates to S. Avery, May 15, 1914, General Correspondence, Avery Papers, University of Nebraska–Lincoln Archives; W. J. Morrill to P. L. Buttrick, May 3, 1915, Forestry Department, Chairman's Correspondence, 1915, University of Nebraska–Lincoln Archives; *Forest Club Annual,* University of Nebraska 2 (1909): 5; 3 (1911): 5; 5 (1913): 62–70, 111–15; 6 (1915): 130–31, 133–37; Sears, "Botanists and Conservation," 360.
94. The best discussion of the later development of the Nebraska National Forest is Pool, "Fifty Years," 139–79. Also, John Clark Hunt, "The Forest That Men Made," *American Forests* 71 (November 1965): 18–21, 46–48; 71 (December 1965): 32–35, 48–50; reprinted by GPO, 1972; Seward D. Smith, "Forestation a Success in the Sand Hills of Nebraska," *Proceedings of the Society of American Foresters* 9 (1914): 388–94; Charles A. Scott, "Foresting the Nebraska Sand-Hills," *Forestry and Irrigation* 9 (1903): 454–57. Also in the summer of 1902 Bessey agreed to start a study of the bull or ponderosa pine for the bureau, emphasizing the extension of the species from the Rockies onto the Great Plains. Bessey to G. B. Sudworth, June 7, 1902, Sudworth to Bessey, June 11, 1902, Bessey to W. O. Jones, September 11, 1902, Bessey to W. L. Hall, October 25, 1902, BP.

## CHAPTER 7

1. Samuel P. Hays, *Conservation and the Gospel of Efficiency: Progressive Conservation Movement* (Cambridge, 1959); Arthur Link and Richard McCormick, *Progressivism* (Arlington Heights, Ill., 1983); Robert H. Wiebe, *The Search for Order, 1877–1920* (New York, 1967).
2. Bessey, "Hear My Children the Words of the Socias," folder 2, Address, folder 1, Presidential Address to the Twenty-Five Year Teachers Club, folder 1, General Botany Notes, Speeches and Articles, BP.
3. E. Crutcher to Bessey, July 17, 1905, BP.
4. Bessey to F. E. Miller, January 8, 1890, BP.
5. H. A. French to Bessey, September 18, 1896, BP.
6. Bessey to F. M. Fitch, January 20, 1914, BP.
7. An example is Bessey to Prof. Wright, April 9, 1892, BP.
8. A. F. Sherril to Bessey, December 29, 1896, O. B. Whitmore to Bessey, March 4, 1898, Bessey to Whitmore, February 10, March 8, 1898, Bessey to *Nebraska State Journal,* September 11, 1906, W. W. Jones to Bessey, December 12, 1907, C. S. Drawbridge to Bessey, April 1909, Bessey to F. M. Fitch, January 20, 1914, BP.
9. Bessey to F. W. Ordway, March 30, 1904, BP.
10. Bessey to F. E. Lloyd, March 29, 1909, BP.
11. Bessey to G. L. Miller, January 22, 1909, BP.
12. Bessey to W. J. Bryan, January 26, 1893, F. V. Coville to J. S. Morton, November 23, 1893, Bessey to J. M. Thurston, March 19, April 3, 1897, Bessey to D. H. Mercer, March 19, 1897, Bessey to J. M. Coulter, April 12, 1897, I. J. Gray to Bessey, August 23,

1898, Bessey to New York *Evening Post,* December 13, 1904, Bessey to J. A. Maguire, May 11, 1909, BP.
   13. Bessey to W. H. Brown, April 9, 1904, BP.
   14. Bessey to D. H. Mercer and E. J. Burkett, December 16, 1901, BP.
   15. Bessey to D. H. Mercer, December 16, 1901, BP.
   16. Bessey to E. M. Pollard, May 5, 1908, BP.
   17. Bessey to W. O. McElroy, October 22, 1892, BP.
   18. Bessey to E. M. Pollard, April 18, 1908, Bessey to E. J. Burkett, April 25, 1908, BP.
   19. Bessey to E. M. Pollard, May 5, 1908, BP.
   20. Bessey to E. J. Burkett, April 25, 1908, BP.
   21. Bessey, "Science and Culture," *Science* 4 (1896): 121–24; Bessey, "Next Steps in Botanical Science," 1–13; Bessey, "Instruction in Pure Science for Agricultural Students," *Proceedings of the Association of American Agricultural Colleges and Experiment Stations* (1914), 213–18.
   22. Bessey to E. J. Burkett, March 5, 1902, BP; Bessey to W. Trelease, March 5, April 24, 1902, N. L. Britton to members of Botanical Society, n.d. [1902], Trelease Papers, Missouri Botanical Garden.
   23. J. M. Cattell to Bessey, September 8, 1902, G. E. MacLean to Bessey, October 27, 1902, R. S. Woodward to Bessey, Bessey to B. M. Davis, October 7, 1902, Bessey to W. F. Ganong, December 6, 1902, Bessey to D. T. MacDougal, April 12, 1907, BP; *Science* 16 (1902): 604–6.
   24. J. P. Norton to Bessey, December 21, 1907, January 29, March 18, 1908, Bessey to Norton, April 21, 1908, J. M. Cattell to Bessey, March 21, 1914, Bessey to Cattell, March 31, April 3, 1914, BP.
   25. Bessey to J. B. Strode, December 15, 1987, BP.
   26. Bessey to D. H. Mercer, February 16, 1901, Bessey to E. J. Burkett, February 16, 1901, March 5. 1902, Bessey to J. H. Millard, December 23, 1901, Bessey to J. Wilson, December 7, 1905, Bessey to A. T. Hadley, April 3, 16, 1909, Bessey to J. A. Maguire, January 12, 1915, BP; Bessey to W. Trelease, March 5, April 24, 1902, N. L. Britton to members of the Botanical Society of America, Trelease Papers, Missouri Botanical Garden. Numerous other examples of interest in government activities are in the Bessey Papers and in Division of Botany and General Correspondence, Division of Botany, USDA, RG 54, National Archives.
   27. Bessey to J. R. Proctor, March 21, 1898, Bessey to J. H. Millard, December 23, 1901, K. Goebel to Bessey, October 1, 1905, BP.
   28. Bessey to J. B. Strode, January 17, 1898, BP.
   29. Bessey to W. H. Brown, April 9, 1904, Bessey to E. M. Pollard, March 7, 1906, Bessey to S. B. Green, October 29, 1908, Bessey to P. W. Ayres, March 16, 1909, Bessey to J. A. Maguire, May 11, 1909, BP.
   30. Bessey to E. M. Pollard, January 8, March 7, 1906, BP.
   31. Bessey to [W. J.] Bryan, December 18, 1894, E. M. Pollard to Bessey, April 21, 30, 1908, Bessey to Pollard, May 5, 1908, BP.
   32. Bessey to W. V. Allen, April 8, 1898, BP.
   33. Bessey to J. M. Thurston, April 23, 1898, BP.
   34. Bessey to J. J. McCarthy, April 22, 1904, BP.
   35. Bessey to F. G. Miller, June 13, 1908, BP.
   36. Bessey to E. M. Pollard, January 28, 1909, BP.
   37. Bessey to E. J. Burkett, February 1, 1907, BP.

38. Bessey to E. J. Burkett, April 25, 1908, Bessey to N. Brown, April 14, 1908, Bessey to E. M. Pollard, April 18, 25, 1908, January 28, 1909, Bessey to C. S. Harrison, May 28, 1908, Bessey to S. B. Green, October 29, 1908, Pollard to Bessey, April 21, 30, 1908, February 6, 16, 1909, BP.

39. Bessey to E. J. Burkett, February 1, 1907, BP.

40. Bessey to F. G. Miller, June 13, 1908, BP.

41. Bessey to E. M. Pollard, May 5, 1908, BP.

42. I. J. Gray to Bessey, October 1, 1908, BP.

43. Bessey to I. J. Gray, October 10, 1908, BP.

44. Bessey to W. L. Hall, June 15, 1903, T. F. Sturgess to Bessey, October 1, 1903, Bessey to Prof. Sargent, December 19, 1903, BP; Bessey, "Notes on the Agriculture of the Caucasus Mountains," *SPAS* (1904), 37–43; Bessey, "Mountain Farming in the Caucasus," *Twentieth Century Farmer* (April 27, 1904), 3; Bessey, "Tea Growing in Transcaucasia," *Pomona College Journal of Economic Botany,* 3 (May 1913): 441–45; Bessey, "Some Foreign Botanical Gardens and Parks," printed in both *Neb Bd Agr* (1903), 82–88, and *Neb Hort Soc* 14 (1904): 157–66; Bessey "Some European Forest Notes," *Forestry Quarterly* 8 (1910): 201–9.

45. Bessey to E. G. Britton, January 28, 1904, Bessey to R. K. Beattie, April 13, 1909, Bessey to F. E. Clements, February 22, 1909, BP; Bessey, "Life in a Seaside Summer School," *Popular Science Monthly* 67 (1905): 80–89; *Science* 30 (1909): 110.

46. Bessey, "Utilizing Wild Species," 111–23; Bessey, "A Preliminary Paper on Drouth Endurance," *Neb Hort Soc* (1913), 245–48.

47. Bessey to H. Bolley, February 12, 1896, BP.

48. Bessey, "The Growing Importance of Plant Physiology in Agricultural Education," *SPAS* (1906), 50–54.

49. Ibid.

50. J. Craig to Bessey, November 15, 21, 1902, Bessey to Craig, November 18, 1902, A. Woods to Bessey, June 15, 1903, BP; Bessey, "Plant Physiology in Agricultural Education," 50–54; Bessey, "Pure Science for Agricultural Students," 213–18.

51. Bessey to H. R. Howe and W. H. Wilson, January 30, 1905, BP; William L. Bowers, *The Country Life Movement in America, 1900–1920* (Port Washington, N.Y., 1974), 4–5, 56–61, 79–82, 109–11, 133; Theodore R. Sizer, *Secondary Schools at the Turn of the Century* (New Haven, 1964), 31–32.

52. Bessey to J. H. Powers, March 25, 1902, Bessey to T. H. Harris, January 23, 1909, F. G. Odell to Bessey, July 22, 1911, Bessey to *Breeder's Gazette,* July 24, 1911, Bessey to W. H. Campbell, July 26, 1911, "The Country Life Problem" and "Nebraska Rural Life Commission (first draft)," folder 2, General Botany Notes, Speeches and Articles, BP; G. W. Hervey, "Nebraska Rural Life Commission," *Twentieth Century Farmer* (July 26, 1911) in File Drawer, C. E. Bessey folder, University of Nebraska–Lincoln Archives; *First Annual Session of the Nebraska Farmers Congress,* 27–29; Bowers, 45–61, 81, 94; Cremin, *Transformation of the School,* 48–50, 82–88.

53. G. B. Sudworth to Bessey, February 3, 1902, Bessey to Sudworth, February 8, 1902, BP; Bessey, "City Trees," *Neb Hort Soc* 16 (1906): 153–56; Bessey, "Physiology of Pruning," 94–98; Bessey, "The Forest Trees of Eastern Nebraska," *Ia Acad Sci* 13 (1906): 75–87.

54. Bessey to G. Pinchot, December 6, 1902, BP.

55. G. B. Sudworth to Bessey, December 12, 1902, BP.

56. Bessey to G. Pinchot, May 20, 1907, BP.

57. F. Clements to Bessey, June 27, 1907, April 7, 1908, BP.

58. Bessey to C. L. Pollard, April 29, 1902, Bessey to L. White, December 1903,

January 7, 1904, White to Bessey, January 12, April 22, 1904, Bessey to E. J. Burkett, January 28, 1904, E. Britton to Bessey, January 28, December 9, 1904, Bessey to J. J. McCarthy, April 22, 1904, Bessey to M. B. Gorham, November 11, 1904, Bessey to S. Coulter, November 18, 1904, W. J. Palmer to Bessey, June 2, 1905, BP; Bessey, "The Wild Flower Preservation Society of America: Report of the President for 1904," *Plant World* 8 (February 1905): 39–42; Bessey, "The Preservation of Wild Flowers," *Science* 16 (1902): 476; ibid. 19 (1904): 165, 176; Tobey, *Saving the Prairies,* 13–14; Walsh, "American Green," 44–47.

59. Weaver and Thiel, 60; Bessey, "A Plan for Completing the State Flora," Science, 29 (1909): 594.

60. Bessey to H. J. Rogers, January 7, 1905, BP; *Science* 8 (1898): 123–26; 24 (1906): 571.

61. Charles R. Barnes, "Suggestions As to Teaching Botany in High Schools," *Science,* o.s., 19 (1892): 91; R. Ellsworth Call, "Natural Science in the High School Course," ibid. 20 (1892): 1; Bruce Fink, "A Mistake in Teaching Botany," ibid. 22 (1893): 217; George H. Hudson, "An Impeachment of 'School Botany,' " ibid. 23 (1894): 103; Katherine E. Golden, "Botany in the Schools," ibid., 119; "Notes and News," ibid., 166; John M. Coulter, "Nature Study and Intellectual Culture," ibid. 4 (1896): 740; Bessey, "High School Botany," ibid. 7 (1898): 266–67; Bessey, "The Conference of Science Teachers in the Trans-Mississippi Educational Convention," ibid. 8 (1898): 123–26; Bessey, Review of textbooks by Liberty Hyde Bailey and Joseph Y. Bergen, ibid. 13 (1901): 859; Review of textbook of Volney M. Spalding, *BG* 18 (1893): 430; Edward A. Krug, *The Shaping of the American High School* (New York, 1964), 35–57; Sizer, xi–xiii, 18–38, 55–72; Tobey, *Saving the Prairies,* 24–32.

62. Sizer, 99–119.

63. Sizer, 218–19, 237–38.

64. Bessey to C. W. Elliott, November 25, 1892, Bessey to W. B. Powell, March 7, April 3, 1893, Bessey to C. B. Scott, March 7, 1893, Bessey to J. M. Coulter, April 4, 1893, Bessey to Prof. Tuttle, [December 1893], BP.

65. C. S. Palmer to Bessey, June 4, July 20, December 21, 1895, J. M. Cattell to Bessey, June 8, 1896, BP; Bessey, "Science and Culture," 121–24.

66. W. S. Jackson to Bessey, January 3, 1896, C. S. Palmer to Bessey, August 17, 1896, May 22, 1897, R. S. Tarr to Bessey, December 12, 1896, Bessey to Tarr, December 15, 1896, Bessey to J. W. [Croltur], February 3, 1897, Bessey to M. A. Bigelow, March 19, 1904, BP; Bessey, "Field Work in Botany in Grammar and High Schools," *Nature-Study Review* 3 (January 1907): 9–16; *Science* 4 (1896): 280; John C. Burnham, *How Superstition Won and Science Lost: Popularizing Science and Health in the United States* (New Brunswick, 1987), 155–58; Cremin, *Transformation of the School,* 75–78.

67. Bessey to Chancellor, January 6, 1892, Bessey to B. Fink, September 12, 1913, BP; Manley, 126–28.

68. Bessey to G. H. Hicks, March 1, 1898, BP; Bessey, "Has the Time Come When None But College Graduates Should Be Elected to High School Positions?" *University Journal* 2 (December 1905): 1–2; Manley, 126–28, 173.

69. Bessey to W. Ganong, March 1, 1901, Ganong to Bessey, March 6, 19, 1901, BP; William F. Ganong, "Advances in Methods of Teaching," 96–100; Ganong, "Note on a Standard College Entrance Option in Botany," ibid. 13 (1901): 611.

70. Bessey to J. M. Coulter, March 27, April 19, 1899, Coulter to Bessey, April 11, 21, May 6, 1899, J. V. Bergen to Bessey, November 6, 1899, Bessey to Bergen, November 9, 1899, Bessey to C. MacMillan, November 17, 1899, BP; *Science* 10 (1899): 857–58, 11 (1900): 466–67; 14 (1901): 494–95.

71. W. Ganong to Bessey, March 19, 1901, BP; *Science* 32 (1910): 671.

72. *Science* 22 (1905): 506–8; 25 (1907): 144–45; 29 (1909): 181; William F. Ganong, "Some Reflections upon Botanical Education in America," ibid. 31 (1910): 321; ibid., 37 (1913): 756–58.

73. *Science* 33 (1911): 633–49 (This included papers by Bessey, "On the Preparation of Botanical Teachers"; O. W. Caldwell, "The Product of Our Botanical Teaching"; F. E. Clements, "Methods of Botanical Teaching"; and commentaries by John M. Coulter and F. C. Newcombe).

74. O. W. Caldwell to Bessey, March 3, 1909, Bessey to Caldwell, March 6, 1909, BP; Bessey, "How Much Plant Pathology Ought a Teacher of Botany to Know?" *Plant World* 8 (August 1905): 189–97; Bessey, "Field Work in Botany," 9–16; Bessey, "A Word As to High School Botany," *University Journal* 4 (October 1907): 11–12; Bessey, "Laboratory Equipment for Botany in the High Schools," ibid. 8 (November 1911): 35–39; Bessey, "The Content of High School Botany," ibid. 8 (February 1912): 89–90.

75. Bessey, "Next Steps in Botanical Science," 2–3.

76. Bessey to J. M. Coulter, May 4, 1899, BP; *Science* 33 (1911): 33, 646; 39 (1914): 72–73, 358–59.

77. Contributions of Department of Botany, [February 1, 1899], Report on Number of Students in Botany, March 15, 1901, Bessey to L. R. Jones, October 21, 1901, Bessey to Chancellor, November 12, 1904, BP.

78. Bessey to H. L. P. Wolcott, September 28, 1894, E. G. Britton to Bessey, July 1, 1896, C. Harrison to Bessey, May 10, 1898, Bessey to F. Detmers, March 28, 1901, A. Ames to Bessey, July 1, 1905, Bessey to Chancellor [1907], E. Field to Bessey, December 5, 1909, Bessey to F. M. Fitch, January 20, 1914, "The Anniversary of a Great Garden," [1914], folder 2, General Botany Notes, Speeches and Articles, BP; Bessey, "Next Steps in Botanical Science," 4–13.

79. Bessey to Chancellor, March 15, 1907, Bessey to J. W. Crabtree, April 7, 1908, F. E. Clements to Bessey, April 28, 1908, Bessey to Clements, [1908], Board of Dean's Report to Chancellor, 1909, folder 6, Botany General Notes, BP; *Science* 22 (1905): 888; 24 (1906): 21–23; 27 (1908): 547–53; 31 (1910): 343–44.

80. E. A. Burnett to Bessey, March 9, 1906, Bessey to W. J. Beal, November 14, 1914, BP.

81. Report of Dean of Industrial College, [December 1898], BP; Manley, 110, 134, 138, 160, 174–75.

82. Manley, 102–3, 175–76.

83. Manley, 148–49, 158–76, 194, 199, 203.

84. S. Avery to Bessey, December 29, 1908, January 9, 1909, Bessey to Avery, January 7, 1909, Bessey to W. J. Beal, November 21, 1914, BP; Bessey, "The Place of the Arts College," *University Journal* 8 (May 1912): 132; *Science* 29 (1909): 613; Manley, 174–76, 188, 194–95, 203.

## CHAPTER 8

1. *Science* 9 (1899): 618; 40 (1914): 750.

2. Peter J. Bowler, *Evolution: The History of an Idea,* rev. ed. (Berkeley, 1983), 246–81; Bowler, *The Non-Darwinian Revolution: Reinterpreting a Historical Myth* (Baltimore, 1988); Ernst Mayr, *The Growth of Biological Thought: Diversity, Evolution, and Inheritance* (Cambridge, 1982), 477–550; Edward J. Pfeifer, "The Genesis of American Neo-Lamarckism," *Isis* 56 (1965): 156–67.

3. *Science,* o.s., 22 (1893): 184; 15 (1902): 991; O. F. Cook, "Natural Selection in Kinetic Evolution," ibid. 19 (1904): 549–50; David Starr Jordan, "The Origin of Species Through Isolation," ibid., 22 (1905): 545, 710, 836.
4. *Science* 15 (1902): 991.
5. *Johnson's Universal Cyclopedia* (1897 edition), J. S. Kingsley, "Darwinism," Vol. 2, 669–70; Kingsley, "Evolution," Vol. 3, 228–34; Kingsley, "Heredity," Vol. 4, 247–48.
6. Ibid., Vol. 4, 248.
7. Liberty Hyde Bailey, "The Plant Individual in the Light of Evolution," *Science* 1 (1895): 281–92; Bailey, "Some Recent Ideas on the Evolution of Plants," ibid. 17 (1903): 441–54 (quoted, 445).
8. Bailey, "Plant Individual," 282.
9. Asa Gray, *Darwiniana: Essays and Reviews Pertaining to Darwinism,* edited by A. Hunter Dupree (Cambridge, Mass., 1963, original 1876), 51–145, 208–32, 293–320; Dupree, *Asa Gray,* 358–59; Bowler, *Evolution,* 260–64; Mayr, *Growth of Biological Thought,* 47–51, 526–31; Ernst Mayr, *Toward a New Philosophy of Biology: Observations of an Evolutionist* (Cambridge, 1988), 241–55.
10. *AN* 21 (1887): 59–60.
11. Bessey to Rev. [H. A. French], July 30, 1892, Bessey to C. Scott, April 26, 1893, Bessey to A. V. Sunderlin, December 17, 1897, W. Boyle to Bessey, April 9, 1907, D. R. Dungan to Bessey, April 24, 1909, Bessey to C. L. Drawbridge, April 1909, BP; Bessey's copy of Asa Gray, *Darwiniana* (New York, 1876), with handwritten marks, underlines, and occasional comments is privately owned by Prof. Roger Sharpe, Department of Biology, University of Nebraska at Omaha.
12. Examples are L. H. Bailey, "Recent Ideas on Evolution," 441–54; *Science* 15 (1902): 721; 20 (1904): 395; Bailey, "Systematic Work and Evolution," ibid. 21 (1905): 532–35; ibid. 34 (1911): 491–93. Paul and Kimmelman, 281–310; Mayr, *Growth of Biological Thought,* 707–27.
13. Bailey, "Recent Ideas on Evolution," 441–45.
14. Bessey to R. A. Emerson, September 8, 1904, BP; Emerson, "Preliminary Account of Variation in Bean Hybrids," *Neb Exp St* (1902), 30–49; Herbert J. Webber, "The Effect of Research in Genetics on the Art of Breeding," *Science* 35 (1912): 597–609; Paul and Kimmelman, 285–86, 294.
15. D. T. MacDougal, "Discontinuous Variation and the Origin of Species," *Science* 21 (1905): 540–43; MacDougal, "Heredity and Environic Forces," ibid. 27 (1908): 121–23; MacDougal, "Organic Response," ibid. 33 (1911): 94–101; Garland E. Allen, "Naturalists and Experimentalists: The Genotype and the Phenotype," *Studies in History of Biology* 3 (1979): 179–209; Allen, *Life Science,* 1–19; Jane Maienschein, "Experimental Biology in Transition: Harrison's Embryology, 1895–1910," ibid. 6 (1983): 107–27; William Coleman, *Biology in the Nineteenth Century: Problems of Form, Function, and Transformation* (New York, 1971), 160–66.
16. Bessey, "The Phyletic Idea in Taxonomy," *Science* 29 (1909): 91.
17. Ibid., 92.
18. Bessey, "The Phylogenetic Taxonomy of Flowering Plants," *Annals of the Missouri Botanical Garden* 2 (1915): 112–15.
19. Charles E. and Ernst A. Bessey, *Essentials of College Botany,* 8th ed. (1914), 157–62; Bessey, *Botany,* 4th ed. (1885), 202–5; Bessey, "Evolution and Classification," *Proceedings of the AAAS* 42 (1893): 237–38.
20. Morton, 293–313, 371–406; Mayr, *Growth of Biological Thought,* 171–220; P. F. Stevens, "Haug and A. P. Candolle: Crystallography, Botanical Systematics, and Comparative Morphology, 1780–1840," *Journal of the History of Biology* 17 (Spring 1984): 49–82.

21. Lester F. Ward, "On the Genealogy of Plants," *AN* 12 (1878): 359–78; Ward, "Classification of Plants," *BG* 8 (1883): 281.
22. Bessey to E. Haeckel, May 8, 1909, BP.
23. *BG* 5 (1880): 74.
24. Bessey to A. J. McClatchie, April 14, 1892, Bessey to N. Britton, April 22, 1892, BP.
25. Bessey to B. Halsted, June 28, 1893, BP.
26. Bessey, "Evolution and Classification," 237–51 (quoted 237, 238, 240, 242).
27. Ibid., 241–48.
28. *BG* 20 (1895): 328; 21 (1896): 307, 375; *Science* 7 (1898): 669; 8 (1898): 136; 12 (1900): 890; 17 (1903): 829–31; 20 (1904): 434–36; 21 (1905): 674–75; 22 (1905): 803–4; 26 (1907): 481.
29. *Science* 4 (1896): 574; 6 (1897): 179.
30. Bessey to N. Britton, November 27, 1894, January 10, 1895, BP.
31. Bessey to B. D. Halsted, November 17, 1897, BP.
32. Bessey to A. J. McClatchie, April 14, 1892, Bessey to M. Fernald, October 24, 1895, BP; *Science* 28 (1908): 688.
33. *BG* 20 (1895): 328.
34. Examples of morphology and classification in the United States: *BG* 10 (1885): 335; Douglas Houghton Campbell, "On the Relationships of the Archegoniata," ibid. 16 (1891): 323–33; Bradley Moore Davis, "The Relationships of Sexual Organs in Plants," ibid. 38 (1904): 241–64; Conway MacMillan, "Current Problems in Plant Morphology: Relationship between Pteridophytes and Gymnosperms," *Science* 7 (1898): 161–64; John M. Coulter, "Development of the Morphological Conceptions," ibid., 20 (1904): 617–24; ibid. 22 (1905): 631; E. C. Jeffrey, "Morphology and Phylogeny," ibid. 23 (1906): 291–97; J. M. Greenman, "Morphology As a Factor in Determining Genetic Relationships," ibid. 41 (1915): 171. For Bessey's early ideas see Chapter 2.
35. Bessey, *Botany,* 4th ed (1885), 204; Bessey, "Phylogeny and Taxonomy of the Angiosperms," *BG* 24 (1897): 157.
36. All by Bessey, "Evolution and Classification," 244–48; "Simplification and Degeneration of Structure in the Angiosperms," *BG* 19 (1894): 369; "Further Studies in the Relationship and Arrangement of the Families of Flowering Plants," ibid., 372–73; "The Point of Divergence of Monocotyledons and Dicotyledons," ibid. 22 (1896): 230–31; "Phylogeny of Angiosperms," 145–78; *Science* 6 (1897): 398.
37. Bessey, "A Synopsis of the Larger Groups of the Vegetable Kingdom," *AN* 28 (1894): 63–65; *BG* 19 (1894): 85.
38. Bessey to A. J. McClatchie, August 23, 1892, BP.
39. Bessey to E. R. Boyer, September 27, 1893, Bessey to J. C. Arthur and N. Britton, November 13, 1893, BP.
40. Bessey, "Taxonomy of Flowering Plants," 118; Lyman Benson, *Plant Classification,* 2nd ed. (Lexington, Mass., 1979), 538.
41. Cited in footnote 36. Also, *BG* 19 (1894): 469; Bessey to Prof. Young, March 11, 1895, Bessey to J. C. Arthur, November 30, 1895, BP; *Johnson's Universal Cyclopedia,* Vol. 8, 453–64; Bessey, *Essentials,* 7th ed. (1896), 320.
42. All by Bessey, "Homologies of Uredineae," 989–96; "The Systematic Arrangement of the Protophyta," *AN* 31 (1897): 63–65; "The Modern Conception of the Structure and Classification of Diatoms," *Transactions of The American Microscopical Society* 21 (1900): 61–85; "The Modern Conception of the Structure and Classification of Desmids," ibid. 22 (1901): 89–97; "The Structure and Classification of the Conjugatae," ibid. 23

(1902): 145–50; "The Structure and Classification of the Phycomycetes," ibid. 24 (1903): 27–54; "Evolution in Microscopic Plants," ibid., 5–12; "The Classification of Protophyta," ibid. 25 (1904): 89–104; "The Structure and Classification of the Lower Green Algae," ibid. 26 (1905): 121–36.
43. Bessey, "Protophyta," 63.
44. Bessey, "Evolution in Microscopic Plants," 5–6.
45. Bessey, "Phyletic Idea," 94.
46. Bessey, "Homologies of Uredineae," 991–94.
47. Bessey, "Protophyta," 63.
48. Bessey, "Lower Green Algae," 134–36.
49. Bessey, "Protophyta," 65; Bessey, "Lower Green Algae," 136.
50. Bessey, "Homologies of Uredineae," 992–94; Bessey, "Phycomycetes," 27–55.
51. Bessey, "Lower Green Algae," 121–37; Bessey, "Conjugatae," 145–50.
52. Bessey, "Evolution in Microscopic Plants," 5.
53. Ibid., 6–11.
54. Ibid., 11.
55. Bessey, "Synopsis of the Conjugate Algae—Zygophyceae," *Transactions of the American Microscopical Society* 33 (1914): 11–49.
56. *BG* 19 (1894): 87; Lucien M. Underwood, "The Evolution of the Hepaticae," ibid., 347–61.
57. Bessey, "Evolution and Classification," 241–44.
58. Bessey, "Point of Divergence," 231.
59. Ibid., 229–32; Bessey to A. J. McClatchie, February 6, 1893, Bessey to E. R. Boyer, September 27, 1893, BP; *Science* o.s., 22 (1893): 105.
60. Bessey, "Evolution and Classification," 244–47; Bessey, "Further Studies of Flowering Plants," 372–73; Bessey, "Phylogeny of Angiosperms," 145–78; Bessey, "Point of Divergence," 229–32; Bessey, "Phyletic Idea," 91–100.
61. *AN* 11 (1877): 61; Lester F. Ward, "Genealogy of Plants," 359–78; Dupree, *Asa Gray,* 394–95.
62. MacMillan, 161–64; John M. Coulter, "The Origin of Gymnosperms and the Seed Habit," *Science* 8 (1898): 377–85; Edward W. Berry, "Recent Discussions of the Origin of Gymnosperms," ibid. 25 (1907): 470–72.
63. D. White to Bessey, November 22, December 4, 1897, Bessey to White, November 30, 1897, BP.
64. Bessey to L. F. Ward, February 4, 1893, BP; Bessey, "Phylogeny of Angiosperms," 145–55.
65. Bessey, "Phyletic Idea," 98.
66. Ibid., 99.
67. Ibid., 91–100; Bessey, "A Synopsis of Plant Phyla," *University Studies,* no. 4, 7 (October 1907): 1–99.
68. Bessey, "Phyletic Idea in Taxonomy," 100.
69. R. J. Harvey-Gibson, *Outlines of the History of Botany* (London, 1919), 256–60.
70. Bessey, *Outline of Plant Phyla* (Lincoln, 1909), 1–20; Bessey, "The Phyla, Classes, and Orders of Plants," *Transactions of the American Microscopical Society* 29 (1910): 85–96; Bessey, "Revisions of Some Plant Phyla," *University Studies,* no. 1, 14 (January 1914): 1–73.
71. Bessey, "Taxonomy of Flowering Plants," 109–64.
72. A summary of his system and its comparison with Engler and Prantl is Benson, 537–41.

73. Volta Torrey, "Battle of Buttercups Won from Germany," Philadelphia *Public Ledger,* May 22, 1932 in General Botany, Bessey's Speeches and Articles, folder of memorabilia, BP.

74. George H. M. Lawrence, *Taxonomy of Vascular Plants* (New York, 1951), 114–40; P. H. Davis and V. H. Heywood, *Principles of Angiosperm Taxonomy* (Princeton, 1963), 58–63; Benson, 537–41; Lloyd H. Swift, *Botanical Classifications: A Comparison of Eight Systems of Angiosperm Classification* (Hamden, Conn., 1974), 7–8; Mayr, *Growth of Biological Thought,* 220–50; Mayr, *New Philosophy of Biology,* 268–88.

75. Arthur Cronquist, *The Evolution and Classification of Flowering Plants,* (Boston, 1968), 52.

76. A. Woods to L. Pammel, April 24, 1906, A. D. Quint to Pammel, April 23, 1906, Pammel to W. B. Allison, April 23, 1906, Pammel to Woods, April 26, 1906, AD-3, Bessey, Pammel Papers.

77. Guy E. Reed to Graduates, n.d. [1915], Senate Committee to Mr. Fitzpatrick, November 15, 1915, BP.

# BIBLIOGRAPHY

## ABBREVIATIONS

Amer Agr—*American Agriculturist for the Farm, Garden, and Household*
AMS—*Transactions of the American Microscopical Society*
AN—*American Naturalist*
BG—*Botanical Gazette*
Ia Acad Sci—*Proceedings of the Iowa Academy of Sciences, 1875–1880*
Ia Agr Soc—*Report of the Iowa State Agricultural Society*
Ia Hort Soc—*Annual Report of the Iowa State Horticultural Society*
Neb Acad Sci—*Publications of the Nebraska Academy of Sciences*
Neb Bd Agr—*Annual Report of the Nebraska State Board of Agriculture*
Neb Exp St—*Annual Report and Bulletins of the Agricultural Experiment Station of Nebraska*
Neb Hort Soc—*Annual Report of the Nebraska State Horticultural Society*
SPAS—*Proceedings of the Society for the Promotion of Agricultural Science*
TCF—*Twentieth Century Farmer*
UJ—*University Journal*

## MANUSCRIPT COLLECTIONS

Atkinson, George. Papers. Department of Manuscripts and University Archives, Cornell University Library, Ithaca, New York.
Avery, Samuel. Papers. University of Nebraska–Lincoln Archives.
Bailey, Liberty Hyde. Papers. Department of Manuscripts and University Archives, Cornell University Library, Ithaca, New York.
Bessey, Charles E. Papers. University of Nebraska–Lincoln Archives.
Britton, Nathaniel. Correspondence. Library of The New York Botanical Garden, Bronx, New York.
Correspondence of State Colleges 1894–1909, Division of Botany, USDA, RG 54, National Archives.
Correspondence of the Forest Service. RG 95, National Archives.
Forestry Department, Chairman's Correspondence. University of Nebraska–Lincoln Archives.

Gray, Asa. Papers. American Philosophical Society Library, Philadelphia.
Gray Herbarium Correspondence. Archives, Harvard University, Cambridge, Massachusetts.
Pammel, L. H. Papers. University Archives, Iowa State University Library, Ames.
Scribner, Frank Lamson. Papers. Manuscripts Division, Library of Congress.
Sem Bot Club Records, Misc. Items (1), University of Nebraska–Lincoln Archives.
Smith, Erwin F. Papers. American Philosophical Society Library, Philadelphia.
State Correspondence 1888–1937, Experiment Stations, USDA, RG 54, National Archives.
Trelease, William. Papers. Department of Manuscripts and University Archives, Cornell University Library, Ithaca, New York.
Trelease, William. Papers. Missouri Botanical Garden Archives and Manuscript Collections, St. Louis.

## DOCUMENTS, JOURNALS, AND NEWSPAPERS

*American Agriculturist*
*American Naturalist*
*Annual Report of the Iowa State Horticultural Society*
*Annual Report of the Nebraska State Board of Agriculture*
*Annual Report of the Nebraska State Horticultural Society*
*Botanical Gazette*
*Botanical Survey of Nebraska,* 1–7 (Lincoln, 1892–1904)
*Breeder's Gazette*
*Daily Nebraska State Journal* (Lincoln)
*First Annual Session of the Nebraska Farmers Congress*
Iowa State Agricultural College
  *Aurora*
  *College Bulletin*
  *Fourth Biennial Report of the Board of Trustees of the Iowa State Agricultural College and Farm to the Governor of Iowa* (1871)
  *Fifth Biennial Report* (1873)
  *Sixth Biennial Report* (1875)
  *Seventh Biennial Report* (1877)
  *Eighth Biennial Report* (1879)
*Iowa State Register* (Des Moines)
*Proceedings of a Convention of Agriculturists Held in the Department of Agriculture,* January 10–18, 1882, USDA, Report no. 22, 3–39.
*Proceedings of the American Association for the Advancement of Science*
*Proceedings of the American Forestry Association*
*Proceedings of the American Pomological Society*

*Proceedings of the Association of American Agricultural Colleges and Experiment Stations*
*Proceedings of the Iowa Academy of Sciences,* 1875–1880
*Proceedings of the Society for the Promotion of Agricultural Science*
*Publications of the Nebraska Academy of Sciences*
*Report of the Bureau of Forestry,* USDA
*Report of the Chief of the Division of Forestry,* USDA (1891–92)
*Report of the Commissioner of Agriculture,* USDA (1885–87)
*Report of the Iowa State Agricultural Society* (1874)
*Report of the Joint Committee of the Fifteenth General Assembly of Iowa Appointed to Visit the State Agricultural College and Farm* (1874)
*Report of the Secretary of Agriculture,* USDA (1889–1915)
Science
*Transactions of the American Microscopical Society*
University of Nebraska
   Annual Report and Bulletins, Agricultural Experiment Station of Nebraska
   Catalogue of 1884–85
   *Contributions of the Botany Department*
   *Forest Club Annual*
   *University Journal*
   *University Studies*

## ARTICLES AND BOOKS

A list of abbreviations appears at the beginning of the bibliography. The published works of Charles E. Bessey are listed in a separate section of the bibliography.

Allen, E. W. "The Society for the Promotion of Agricultural Science: Its Present and Its Future." *SPAS* (1920), 96–99.

Allen, Garland E. *Life Science in the Twentieth Century.* New York: John Wiley, 1975.

———. "Naturalists and Experimentalists: The Genotype and the Phenotype." *Studies in History of Biology,* 3 (1979): 179–209.

Alvord, Benjamin. "On the Compass Plant." *AN* 16 (1882): 633–35.

Appel, Toby A. "Organizing Biology: The American Society of Naturalists and Its 'Affiliated Societies,' 1883–1923." In Ronald Rainger, Keith R. Benson, and Jane Maienschein, eds. *The American Development of Biology.* Philadelphia: Univ. of Pennsylvania Press, 1988, 87–120.

Arthur, J. C. "Botanists and Botanizing at Minneapolis." *BG* 8 (1883): 296.

———. "Preliminary List of Iowa Uredineae." *College Bulletin,* Iowa State Agricultural College (1884), 151.

———. "Some Botanical Laboratories of the United States." *BG* 10 (1885): 395–406.

Bailey, Joseph Cannon. *Seaman A. Knapp: Schoolmaster of American Agriculture.* New York: Columbia University Press, 1945.
Bailey, Liberty Hyde. "The Plant Individual in the Light of Evolution." *Science* 1 (1895): 281–92.
_____. "Some Recent Ideas on the Evolution of Plants." *Science* 17 (1903): 441–54.
_____. "Systematic Work and Evolution." *Science* 21 (1905): 540–43.
Baker, Kenneth. "Fire Blight of Pome Fruits: The Genesis of the Concept That Bacteria Can Be Pathogenic to Plants." *Hilgardia* no. 18, 40 (July 1971): 6–33.
Barnes, Charles R. "Suggestions As to Teaching Botany in High Schools." *Science*, o.s., 19 (1892): 91.
Bates, J. M. "The Grasses of Northwestern Nebraska." *Neb Bd Agr* (1891), 130–34.
Beal, W. J. "The Society for the Promotion of Agricultural Science." *SPAS* (1907), 27–30.
_____. *History of the Michigan Agricultural College and Biographical Sketches of Trustees and Professors.* East Lansing: Agricultural College, 1915.
Beardsley, Edward H. *The Rise of the American Chemistry Profession, 1850–1900.* University of Florida Monographs: Social Sciences, no. 23 (Summer 1964).
Bell, P. R., ed. *Darwin's Biological Work: Some Aspects Reconsidered.* New York: John Wiley and Sons, 1964.
Benjamin, Marcus. "Charles Edwin Bessey." *UJ* 8 (Jan. 1912): 66.
Benson, Lyman. *Plant Classification.* 2nd ed. Lexington, Mass.: D. C. Heath, 1979.
Berry, Edward W. "Recent Discussions of the Origin of Gymnosperms." *Science*, 25 (1907): 470–72.
Bledstein, Burton J. *The Culture of Professionalism: The Middle Class and the Development of Higher Education in America.* New York: W. W. Norton, 1976.
Bowers, William L. *The Country Life Movement in America, 1900–1920.* Port Washington, N.Y.: Kennikat Press, 1974.
Bowler, Peter J. *Evolution: The History of an Idea.* rev. ed. Berkeley: Univ. of California Press, 1983.
_____. *The Non-Darwinian Revolution: Reinterpreting a Historical Myth.* Baltimore: Johns Hopkins Univ. Press, 1988.
Britton, Nathaniel Lord, and Addison Brown. *An Illustrated Flora of the Nothern United States, Canada, and the British Possessions.* New York: C. Scribner's Sons, 1896.
Bruce, Robert V. *The Launching of Modern American Science, 1846–1876.* Ithaca: Cornell Univ. Press, 1987.
Budd, J. L. "Experimental Horticulture." *Ia Hort Soc* 16 (1881): 217–21.
Burnham, John C. *How Superstition Won and Science Lost: Popularizing Science and Health in the United States.* New Brunswick: Rutgers Univ. Press, 1987.

Call, R. Ellsworth. "Natural Science in the High School Course." *Science,* o.s., 20 (1892): 1.
Campbell, Douglas Houghton. "On the Relationships of the Archegoniata." *BG* 16 (1891): 323–33.
Cittadino, Eugene. "Ecology and the Professionalization of Botany in America, 1890–1905." *Studies in History of Biology* 4 (1980): 171–98.
Clinton, G. P. "Botany in Relation to Agriculture." *Science* 43 (1916): 1–13.
Cohen, I. Bernard. "Science in America: The Nineteenth Century." In Arthur M. Schlesinger, Jr., and Morton White, eds. *Paths of American Thought.* Boston: Houghton Mifflin, 1963, 181–88.
Coleman, William. *Biology in the Nineteenth Century: Problems of Form, Function, and Transformation.* New York: John Wiley, 1971.
Cook, O. F. "Natural Selection in Kinetic Evolution." *Science* 19 (1904): 549–50.
Coulter, John M. "Laboratory Appliances." *BG* 10 (1885): 409.
———. "Laboratory Courses of Instruction." *BG* 10 (1885): 417.
———. "Nature Study and Intellectual Culture." *Science* 4 (1896): 740.
———. "The Origin of Gymnosperms and the Seed Habit." *Science* 8 (1898): 377–85.
———. *Plant Relations: A First Book of Botany.* New York: D. Appleton, 1900.
———. "Development of the Morphological Conceptions." *Science* 20 (1904): 617–24.
Crawford, Robert Platt. *These Fifty Years: A History of the College of Agriculture of the University of Nebraska.* Lincoln: College of Agriculture, 1925.
Cremin, Lawrence A. *Transformation of the School: Progressivism in American Education, 1876–1957.* New York: Alfred A. Knopf, 1964.
———. *American Education: The National Experience, 1783–1876.* New York: Harper and Row, 1980.
———. *American Education: The Metropolitan Experience, 1876–1980.* New York: Harper and Row, 1988.
Cronquist, Arthur. *The Evolution and Classification of Flowering Plants.* Boston: Houghton Mifflin, 1968.
Daniels, George H. "The Process of Professionalization in American Science: The Emergent Period, 1820–1860." *Isis* 58 (1967): 151–66.
———. "The Pure-Science Ideal and Democratic Culture." *Science* 156 (1967): 1699–1701.
Davis, Bradley Moore. "The Relationships of Sexual Organs in Plants." *BG* 38 (1904): 241–64.
Davis, P. H., and V. H. Heywood. *Principles of Angiosperm Taxonomy.* Princeton, N.J.: Van Nostrand, 1963.
de Beer, Gavin. *Charles Darwin: A Scientific Biography.* Garden City, N.Y.: Doubleday, 1965.
Dupree, A. Hunter. *Science in the Federal Government: A History of Policies and Activities to 1940.* New York: Harper and Row, 1957.
———. *Asa Gray, 1810–1888.* 1959. Reprint. Cambridge: Belknap Press, Atheneum, 1968.
Eddy, Edward Danforth, Jr. *Colleges for Our Land and Time: The Land-Grant*

*Idea in American Education.* New York: Harper, 1957.
Emerson, R. A. "Preliminary Account of Variation in Bean Hybrids." *Neb Exp St* (1902), 30–49.
Emmons, David M. "Theories of Increased Rainfall and the Timber Culture Act of 1873." *Forest History* 15 (1971): 6–14.
Farlow, William G. "Additions to the Peronosporeae of the U.S." *BG* 8 (1883): 37.
_____. "Notes on Some Ustilagineae of the U.S." *BG* 8 (1883): 271–77.
Fink, Bruce. "A Mistake in Teaching Botany." *Science,* o.s., 22 (1893): 217.
Furnas, Robert W. "Tree Growth on the Plains." *Ia Hort Soc* 20 (1886): 380–81.
Galloway, Beverly T. "The Twentieth Century in Botany." *Science* 19 (1904): 11–18.
Ganong, W. F. "Advances in Methods of Teaching: Botany." *Science* 9 (1899): 96–100.
_____. "Some Reflections upon Botanical Education in America." *Science* 31 (1910): 321.
Gilman, Daniel Coit, ed. *Recent Information Respecting Agricultural Education Elsewhere.* Berkeley: Univ. of California, 1874.
Gleason, H. A. "Twenty-Five Years of Ecology, 1910–1935." *Brooklyn Botanic Garden Memoirs* 4 (1936): 41–49.
Golden, Katherine E. "Botany in the Schools." *Science* o.s., 23 (1894): 119.
Gray, Asa. *Gray's Lessons in Botany: The Element of Botany for Beginners and for Schools.* New York: American Book, 1887.
_____. *Darwiniana: Essays and Reviews Pertaining to Darwinism.* Edited by A. Hunter Dupree. Cambridge, Mass.: Belknap Press, 1963.
_____. *Gray's School and Field Botany,* rev. ed. New York: American Book, 1887.
Greenman, J. M. "Morphology As a Factor in Determining Genetic Relationships." *Science* 41 (1915): 171.
Guralnick, Stanley M. "Sources of Misconception on the Role of Science in the Nineteenth-Century American College." *Isis* 65 (Sept. 1974): 352–66.
_____. *Science and the Ante-Bellum American College.* Philadelphia: American Philosophical Society, 1975.
Hall, William L. "The Investigation Now Being Made in Nebraska by the U.S. Bureau of Forestry." *Forester* 7 (1901): 188–93.
Halsted, Byron D. "The Society's Progress." *SPAS* (1898), 20–42.
Harvey-Gibson, R.J. *Outlines of the History of Botany.* London: A & C Black, 1919.
Hay, O. P. "An Examination of Prof. Leo Lesquereux's Theory of the Origin and Formation of Prairies." *Ia Hort Soc* 12 (1878): 299.
Hays, Samuel P. *Covservation and the Gospel of Efficiency: Progressive Conservation Movement.* Cambridge: Harvard Univ. Press, 1959.
Hudson, George H. "An Impeachment of 'School Botany.'" *Science,* o.s., 23 (1894): 103.
Hunt, John Clark. "The Forest That Men Made." *American Forests* 71 (Nov. 1965): 18–21, 46–48; ibid. (Dec. 1965): 32–35, 48–50.

Hussong, E. M. "The Yellow Pine in the Republican Valley." *Neb Acad Sci* 5 (1894): 7–8.
Jeffrey, E. C. "Morphology and Phylogeny." *Science* 23 (1906): 291–97.
Jordan, David Starr. "The Origin of Species Through Isolation." *Science* 22 (1905): 545, 710, 836.
Jordan, W. H. "The Promotion of Agricultural Science." *SPAS* (1902), 22–33.
Kevles, Daniel J. "American Science." In Nathan O. Hatch (ed.), *The Professions in American History.* Notre Dame: Univ. of Notre Dame Press, 1988, 107–25.
Kingsley, J. S. "The Hat Creek (Nebraska) Bad Lands." *AN* 25 (1891): 963–64.
———. "Darwinism." *Johnson's Universal Cyclopedia,* 1897 ed., Vol. 2, 669–70; "Evolution," Vol. 3, 228–34; "Heredity," Vol. 4, 247–48.
Knapp, Seaman A. "The Agricultural College, As It Is, And As It Ought to Be." *Ia Agr Soc* (1881), 243–53.
Krug, Edward A. *The Shaping of the American High School.* New York: Harper and Row, 1964.
Kuhn, Madison. *Michigan State: The First Hundred Years.* East Lansing: Michigan State Univ. Press, 1955.
Kutzleb, Charles R. "Can Forests Bring Rain to the Plains?" *Forest History* 15 (1971): 14–21.
Lawrence, George H. M. *Taxonomy of Vascular Plants.* New York: MacMillan, 1951.
Layton, Edwin. "Mirror-Image Twins: The Communities of Science and Technology in 19th-Century America." *Technology and Culture* 12 (Oct. 1971): 562–80.
Link, Arthur, and Richard McCormick. *Progressivism.* Arlington Heights, Ill.: Harlan Davidson, 1983.
McAfee, H. H. "Report of Committee on Forestry." *Ia Hort Soc* 9 (1874): 135–47.
———. "Pre-Natal Influences." *Proceedings of the Sixth Annual Meeting of the Eastern Iowa Horticultural Society,* bound with *Ia Hort Soc* 10 (1875): 349–50.
MacDougal, D. T. "Discontinuous Variation and the Origin of Species." *Science* 21 (1905): 540–43.
———. "Heredity and Environic Forces." *Science* 27 (1908): 121–23.
———. "Organic Response." *Science* 33 (1911): 94–101.
MacMillan, Conway. "Current Problems in Plant Morphology: Relationship between Pteridophytes and Gymnosperms." *Science* 7 (1898): 161–64.
McNab, William Ramsey. *Botany: Outlines of Morphology, Physiology and Classification of Plants.* Revised for American Students by Charles E. Bessey, New York: Henry Holt, 1881.
Macomber, J. K. "The Relation of Forest to Climate." *Ia Hort Soc* 10 (1875): 241–47.
———. "Adaptability of Prairie Soils for Timber Growth." *Ia Hort Soc* 14 (1879): 292–97.
Maienschein, Jane. "Experimental Biology in Transition: Harrison's Embryol-

ogy, 1895–1910." *Studies in History of Biology* 6 (1983): 107–27.
Manley, Robert N. *Centennial History of the University of Nebraska: A Frontier History, (1869–1919).* Lincoln: Univ. of Nebraska Press, 1969.
Marcus, Alan I. *Agricultural Science and the Quest for Legitimacy.* Ames: Iowa State Univ. Press, 1985.
Mayr, Ernst. *The Growth of Biological Thought: Diversity, Evolution, and Inheritance.* Cambridge: Belknap Press, 1982.
———. *Toward a New Philosophy of Biology: Observations of an Evolutionist.* Cambridge: Belknap Press, 1988.
Morton, A. G. *History of Botanical Science.* London: Academic Press, 1981.
Overfield, Richard A. "Hog Cholera, Texas Fever, and Frank S. Billings: An Episode in Nebraska Veterinary Science." *Nebraska History* 57 (Spring 1976): 99–128.
Pammel, L. H. "Dr. Charles Edwin Bessey," *Prominent Men I Have Met.* Ames: n.p., 1928, 3–5.
Parry, C. C. "Botany and Horticulture." *Ia Hort Soc* 8 (1874): 56.
———. "Notes on Rocky Mountain Conifers As Adapted to Cultivation in the Central Northwestern States." *Ia Hort Soc* 14 (1879): 343–47.
Paul, Diane B., and Barbara A. Kimmelman. "Mendel in America: Theory and Practice, 1900–1919." In Ronald Rainger, Keith R. Benson, and Jane Maienschein, eds. *The American Development of Biology.* Philadelphia: Univ. of Pennsylvania Press, 1988, 281–310.
Pedersen-Bjergaard, J. "Forest Planting on the Sands of Denmark." *Amer Agr* (1893), 19.
Pfeifer, Edward J. "The Genesis of American Neo-Lamarckism." *Isis* 56 (1965): 156–67.
Pinchot, Gifford. "The Immediate Future in Forest Work." *Forestry and Irrigation* 8 (1902): 18–21.
Pool, Raymond J. "A Brief Sketch of the Life and Work of Charles Edwin Bessey." *American Journal of Botany* 2 (Dec. 1915): 505–17.
———. "The Evolution and Differentiation of Laboratory Teaching in the Botanical Sciences." *Iowa State Journal of Science* 3 (Jan. 1935): 238–39.
———. "Fifty Years on the Nebraska National Forest." *Nebraska History* 34 (1953): 139–79.
Pound, Roscoe. "Progress of the Botanical Survey of Nebraska." *Neb Acad Sci* 4 (1894): 7–8.
———. "The Plant-Geography of Germany." *AN* 30 (1896): 465–68.
Pound, Roscoe, and Frederic E. Clements. *The Phytogeography of Nebraska, A General Survey.* Lincoln: Seminar, 1898.
———. "The Vegetation Regions of the Prairie Province." *BG* 25 (1898): 381–94.
Reingold, Nathan. "Alexander Dallas Bache: Science and Technology in the American Idiom." *Technology and Culture* 11 (Apr. 1970): 163–77.
Roberts, Isaac Phillips. *Autobiography of a Farm Boy.* Ithaca: Cornell Univ. Press, 1946.

———. "The Promotion of Agricultural Science." *SPAS* (1897), 82–85.
Rodgers, Andrew Denny III. *American Botany, 1873–1892: Decades of Transition.* 1944. Reprint. New Brunswick, N.J.: Princeton Univ. Press, Hafner, 1968.
———. *John Merle Coulter: Missionary in Science.* Princeton, N.J.: Princeton Univ. Press, 1944.
———. *Bernhard Eduard Fernow: A Story of North American Forestry.* New Brunswick, N.J.: Princeton Univ. Press, 1951.
———. *Erwin Frink Smith: A Story of North American Plant Pathology.* Philadelphia: American Philosophical Society, 1952.
Rosenberg, Charles E. "The Adams Act: Politics and the Cause of Scientific Research." *Agricultural History* 38 (Jan. 1964): 3–12.
———. "Science, Technology, and Economic Growth: The Case of the Agricultural Experiment Station Scientist, 1875–1914." *Agricultural History* 45 (Jan. 1971): 1–20.
———. "Science and Social Values in Nineteenth-Century America: A Case Study in the Growth of Scientific Institutions." In *No Other Gods: On Science and American Social Thought.* Baltimore: Johns Hopkins Univ. Press, 1978, 135–52.
Ross, Earle D. *Democracy's College: The Land-Grant Movement in the Formative Stage.* Ames: Iowa State College Press, 1942.
———. *A History of Iowa State College of Agriculture and Mechanic Arts.* Ames: Iowa State College Press, 1942.
Rossiter, Margaret W. *The Emergence of Agricultural Science: Justus Liebig and the Americans, 1840–1880.* New Haven: Yale Univ. Press, 1975.
———. "The Organization of the Agricultural Sciences." In Alexandra Oleson and John Voss, eds. *The Organization of Knowledge in Modern America, 1860–1920.* Baltimore: Johns Hopkins Univ. Press, 1979, 211–48.
Rothrock, Joseph T. "Home and Foreign Modes of Teaching Botany, III." *BG* 6 (1881): 233.
Rudolph, Emanuel D. "The Botanical Textbook in Nineteenth-Century America As a Refection of Botanical and Cultural Trends." Paper presented to the Midwest Junto for the History of Science, 1975.
Rudolph, Frederick. *The American College and University: A History.* New York: Knopf, 1962.
Rydberg, Per Axel. "Some Grasses of Southwestern Nebraska." *Neb Bd Agr* (1891), 134–37.
———. "The Flora of the High Nebraska Plains." *AN* 25 (1891): 485.
———. "Flora of the Sand Hills of Nebraska." *Contributions from the United States National Herbarium,* no. 3, 3 (1895): 145–46.
Scott, Charles A. "Foresting the Nebraska Sand-Hills." *Forestry and Irrigation* 9 (1903): 454–57.
Sears, Paul B. "Botanists and the Conservation of Natural Resources." In William Campbell Steere, ed. *Fifty Years of Botany.* New York: McGraw Hill, 1958, 359–66.

———. "Plant Ecology." In Joseph Ewan, ed. *A Short History of Botany in the United States*. New York: Hafner, 1969, 124–31.
Shils, Edward. "The Order of Learning in the United States: The Ascendancy of the University." In Alexandra Oleson and John Voss, eds. *The Organization of Knowledge in Modern America, 1860–1920*. Baltimore: Johns Hopkins Univ. Press, 1979, 19–47.
Sizer, Theodore R. *Secondary Schools at the Turn of the Century*. New Haven: Yale Univ. Press, 1964.
Smith, Jared. "The Grasses of the Sand Hills of Northern Nebraska." *Neb Bd Agr* (1891), 280–91.
Smith, Seward D. "Forestation a Success in the Sand Hills of Nebraska." *Proceedings of the Society of American Foresters* 9 (1914): 388–94.
Stevens, P. F. "Haug and A.-P. de Candolle: Crystallography, Botanical Systematics, and Comparative Morphology, 1780–1840." *Journal of the History of Biology* 17 (Spring 1984): 49–82.
Svoboda, Joseph G., and Patricia Churray, eds. "Guide and Index to the Microfilm Edition of the Charles E. Bessey Papers (1865–1915)." *University of Nebraska Studies*, n.s., no. 67 (1984).
Swezey, G. D. "Additions to the Flora of Nebraska." *Neb Acad Sci* 2 (1892): 16–17.
Thorne, Charles E. "Our Place in the Sun." *SPAS* (1916), 6–14.
Tippo, Oswald. "The Early History of the Botanical Society of America." In William Campbell Steere, ed. *Fifty Years of Botany*. New York: McGraw Hill, 1–13.
Tobey, Ronald C. "Theoretical Science and Technology in American Ecology." *Technology and Culture* 17 (1976): 718–28.
———. *Saving the Prairies: The Life Cycle of the Founding School of American Plant Ecology, 1895–1955*. Berkeley: Univ. of California Press, 1981.
Todd, J. E. "Notes on the Distribution of Timber in South-Western Iowa, with Inferences Concerning the Origin of Prairies." *Ia Hort Soc* 12 (1878): 91.
———. "The Relation of Forests to Rainfall." *Ia Hort Soc* 13 (1878): 112–18.
True, Alfred Charles. *A History of Agricultural Education in the United States, 1785–1925*. Washington, D.C.: Government Printing Office, 1929.
———. *A History of Agricultural Experimentation and Research in the United States, 1607–1925*. Misc. Publication No. 251, USDA. Washington, D.C.: Government Printing Office, 1937.
Underwood, Lucien M. "The Evolution of the Hepaticea." *BG* 19 (1894): 347–61.
Veysey, Laurence R. *The Emergence of the American University*. Chicago: Univ. of Chicago Press, 1965.
Walsh, Thomas R. "Charles E. Bessey and the Transformation of the Industrial College." *Nebraska History* 52 (1971): 383–409.
———. "The American Green of Charles Bessey." *Nebraska History* 53 (1972): 35–57.
———. "Charles E. Bessey: Land-Grant College Professor." Ph.D. diss., University of Nebraska–Lincoln, 1972.

Ward, Lester F. "On the Genealogy of Plants." *AN* 12 (1878): 359–78.
_____. "Classification of Plants." *BG* 8 (1883): 281.
Weaver, J. E. and Albert F. Thiel. "Ecological Studies in the Tension Zone between Prairie and Woodland." *Botanical Survey of Nebraska,* n.s., no. 1 (April 1917).
Webber, Herbert J. "The Flora of Central Nebraska." *AN* 23 (1889): 633–36.
_____. "The Fresh-Water Algae of the Plains." *AN* 23 (1889): 1011–13.
_____. "Appendix to the Catalogue of the Flora of Nebraska." *Neb Bd Agr* (1892), 45–53.
_____. "The Effect of Research in Genetics on the Art of Breeding." *Science* 35 (1912): 597–609.
Weiner, Charles. "Science and Higher Education." In David D. Van Tassel and Michael G. Hall, eds. *Science and Society in the United States.* Homewood, Ill.: Dorsey Press, 1966, 163–89.
Welch, A. S. "The Relation of the Agricultural College to Horticulture." *Ia Hort Soc* 9 (1875): 148–60.
_____. "Science with Practice in Education." *Ia Hort Soc* 16 (1881): 153–60.
_____. "The Effect of Horticulture on the Horticulturist." *Ia Hort Soc* 17 (1882): 153–60.
White, C. A. "Adaptability of the Prairies for Artifical Forestry." *Science,* o.s., 3 (1884): 438–42.
Whitney, J. D. "Plain, Prairie, and Forest." *AN* 10 (1876): 577.
Wiebe, Robert H. *The Search for Order, 1877–1920.* New York: Hill and Wang, 1967.
Williams, Thomas A. "Notes on Nebraska Lichens." *AN* 23 (1889): 161.
_____. "Notes on the Canyon Flora of Northwest Nebraska." *AN* 24 (1890): 779.
_____. "Notes on the Flora of Western South Dakota." *AN* 26 (1892): 60, 253.
Woods, Albert. "A Few Notes on the Grasses of the 'Bad Lands' [of Nebraska]." *Neb Bd Agr* (1892), 291–93.

## CHARLES E. BESSEY:
## BOOKS, ARTICLES, AND SELECTED NOTES AND PAPERS

The following entries are listed in approximate chronological order of publication. A list of abbreviations appears at the beginning of the bibliography.

"Injurious Insects." *Ia Hort Soc* 6 (1871): 162–67.
"Contributions to the Flora of Iowa." *Fourth Biennial Report, Iowa State Agricultural College* (1871), 89–127.
"*Lemna polyrrhiza.*" *AN* 6 (1872): 636.
"Sensitive Stamens in Portulaca." *AN* 7 (1873): 464–65.
"Microscopic Examinations at the Agricultural College upon Leaf and Cell Growth." *Ia Hort Soc* 8 (1873): 22.
"Report on Botany, Zoology, and Horticulture." *Fifth Biennial Report, Iowa*

*State Agricultural College* (1873), 91–96.
"Report on Insects Injurious to the Plants and Animals of the Farm." *Ia Agr Soc* (1874): 232–53.
"Double Thalictrum." *AN* 8 (1874): 499.
"*Adoxa Moschatellina L.* in Iowa." *AN* 8 (1874): 690.
"*Botrychium Lunaria Swartz.*" *AN* 8 (1874): 691.
"Choosing a Microscope." *Aurora* (April 1874), 7–8.
"Using a Microscope." *Aurora* (May 1874), 7–8.
"On Injurious Fungi." *Sixth Biennial Report, Iowa State Agricultural College* (1875), 128–33.
"Fungus on Cottonwood." *Ia Hort Soc* 10 (1875): 86.
"A New Fungus on the Ash." *Ia Hort Soc* 10 (1875): 87.
"Some Effects of Low Temperature upon Plants, with a Review of the Nature of Protoplasm." *Ia Hort Soc* 11 (1876): 88–93.
"The Climatic Adaptation of Plants." *Ia Hort Soc* 12 (1877): 86–97.
"A Preliminary Catalogue of the Orthoptera of Iowa." *Seventh Biennial Report, Iowa State Agricultural College* (1877), 205. (also in *Ia Acad Sci* [1876], 8)
"Observations on *Silphium laciniatum,* the So-Called Compass Plant." *AN* 11 (1877): 486–89. (also in *Ia Acad Sci* [1876], 10)
"On Injurious Fungi." *Seventh Biennial Report, Iowa State Agricultural College* (1877), 185–204.
"Further Observations upon *Silphium laciniatum.*" *Ia Acad Sci* (1877), 13.
"*Silphium laciniatum.*" *AN* 11 (1877): 564.
"On a Scientific Course of Study." *Aurora* (May 1877), 5–6; (June 1877), 5; (July 1877), 6.
"Botany in Its Relation to Horticulture." *Ia Hort Soc* 13 (1878): 222–28.
*Geography of Iowa, A Supplement to the Eclectic Series of Geographies.* Cincinnati: Van Antwerp, Bragg, 1878.
"Botanical Aspects of Apple Blight." *Eighth Biennial Report, Iowa State Agricultural College* (1879), 106–7.
"Popular Contributions from the Botanical Laboratory." *Eighth Biennial Report, Iowa State Agricultural College* (1879), 100–14.
"Synopsis of a Lecture Upon the Leaf: Its Structure, Functions and Climatic Modification." *Ia Hort Soc* 14 (1879): 130–34.
"Contributions to the Bryology of Iowa." *Ia Acad Sci* (1876), 7.
"A Preliminary Catalogue of the Lichens of Iowa." *Ia Acad Sci* (1876), 8.
"Notes on the Colors of the Native Wild Flowers of Iowa." *Ia Acad Sci* (1876), 8.
"A Case of Natural Selection." *Ia Acad Sci,* (1876), 10.
"Some Observations upon the Growth of Plants, Made by Means of the Arc-Indicator." *Ia Acad Sci* (1876), 10.
"Notes on the Dimorphism of *Oxalis violacea.*" *Ia Acad Sci* (1877) 11.
"Note on the Forms of the Flowers of *Lithospermum longiflorum.*" *Ia Acad Sci* (1877), 13.
"On the Affinities of the Uredineae." *Ia Acad Sci* (1878), 14.
"On a Botanical Map of the United States." *Ia Acad Sci* (1879), 18.
"Sketch of a Natural Arrangement of Plants." *Ia Acad Sci* (1879), 18.

"Sketch of a Course of Laboratory Practice in Higher Botany." *Ia Acad Sci* (1879), 20.
"A Classification of the Tissues of Plants." *Ia Acad Sci* (1879), 20.
"The Morphology of the Iris Leaf." *Ia Acad Sci* (1880), 22.
"A Genealogical Tree of the Vegetable Kingdom." *Ia Acad Sci* (1880), 24.
"Notice of a Simple Dendrometer." *Ia Acad Sci* (1880), 24.
"The Supposed Dimorphism of *Lithospermum longiflorum*." *AN* 14 (1880): 417–21.
"A Sketch of the Progress of Botany in the United States in the Year 1879." *AN* 14 (1880): 862–70.
"The Flower: Its Structural and Functional Meaning." *Ia Hort Soc* 15 (1880): 174–82.
*Botany for High Schools and Colleges.* Philadelphia: Henry Holt, 1880.
"A Sketch of the Progress of Botany in the United States in 1880." *AN* 15 (1881): 947–55.
"Insect-Destroying Fungi." *AN* 15 (1881): 52–53.
"The Fungi Which Produce Mildew on Cotton Goods." *AN* 15 (1881): 132.
"An Easily Made Observation." *BG* 6 (1881): 172.
"The Asparagus for Histological Study." *BG* 6 (1881): 294–95.
"The Superabundance of Pollen in Indian Corn." *AN* 15 (1881): 1000.
"The Diseases of Plants." *Ia Hort Soc* 17 (1882): 85–98.
"On Parasitic and Other Fungi." *Ia Hort Soc* 17 (1882): 280–84.
"Modern Botany and Mr. Darwin." *AN* 16 (1882): 507–8.
"The Plague of Cut-Worms." reprinted from the *Chicago Herald* in the *Ia Hort Soc* 16 (1882): 332–33.
"Some Observations on the Action of Frost upon Leaf-Cells." *Proceedings of the AAAS* 31 (1882): 464–65.
"Two Parasitic Fungi." *Proceedings of the Eastern Iowa Horticultural Society*, bound with *Ia Hort Soc* 18 (1883): 521–23.
"The Functions of the Leaf." *Ia Hort Soc* 18 (1883): 195–200.
"A New Species of Insect-Destroying Fungus." *AN* 17 (1883): 1280–81.
"Remarkable Fall of Pine Pollen." *AN* 17 (1883): 658.
"The Bearing of Plowright's Discovery As to the Germination of Rust-Spores upon the Problem of Wheat-Growing in the North-West." *SPAS* (1883), 31.
*The Essentials of Botany.* The American Science Series, Briefer Course, vol. V. New York: Henry Holt, 1884.
"Popular Descriptions of Some Harmful Plants." *College Bulletin* (1884), 110–32.
"Preliminary List of Cryptogams." *College Bulletin* (1884), 133–50.
"The Condition of the Living Plant in Winter." *Ia Hort Soc* 19 (1884): 200–6.
"Science and Practice." *Daily State Journal* (Lincoln), September 17, 1884.
"Hybridism in Spirogyra." *AN* 18 (1884): 67–68.
"Glands on a Grass." *AN* 18 (1884): 420–21.
"Sexuality in Zygnemaceae." *AN* 18 (1884): 421–22.
"A Government Duty." *AN* 18 (1884): 543.
"Structure of the Fruit of Porcupine Grass." *AN* 18 (1884): 930–31.

"Adventitious Inflorescence of *Cuscuta glomerata.*" *AN* 18 (1884): 1145–47.
"Curvature of the Stems of Conifers." *BG* 9 (1884): 156.
"Mode of Opening of the Flowers of *Desmodium sessilifolium.*" *BG* 9 (1884): 156.
"A Point in the Structure of the Sterile Flowers of Silphium." *BG* 9 (1884): 156.
"Corn Smut." *Students' Farm Journal* (Ames), 1 (Nov 1, 1884).
"Injurious Fungi, in Their Relation to the Diseases of Plants." *Proceedings of the American Pomological Society* (1885), 35–43.
"The Question of Bisexuality in the Pond-Scums." *BG* 10 (1885): 334.
"Further Observations on the Adventitious Inflorescence of *Cuscuta glomerata.*" *BG* 10 (1885): 334.
"The Study of Parasitic Fungi." *AN* 19 (1885): 170–71.
"Plant Migrations." *AN* 19 (1885): 398–99.
"The Study of Liverworts in North America." *AN* 19 (1885): 604.
"The Opening of the Flowers of *Desmodium sessilifolium.*" *AN* 19 (1885): 711–13.
"Attempted Hybridization Between Pond-Scums of Different Genera." *AN* 19 (1885): 800–2.
"The Movement of Protoplasm in the Styles of Indian Corn." *AN* 19 (1885): 888.
"The Abundance of Ash Rust." *AN* 19 (1885): 886–87.
"The Demands Made by Agriculture upon the Science of Botany." *SPAS* (1885), 16–18.
"The Grasses and Forage Plants of Nebraska." *Neb Bd Agr* (1886), 204–37.
"Pear Blight Bacteria and the Horticulturists." *AN* 20 (1886): 166.
"Books on Fungi." *AN* 20 (1886): 645.
"The Rust of the Ash Tree." *AN* 20 (1886): 806.
"The Roughness of Certain Uredospores." *AN* 20 (1886): 1053.
"Specimens and Specimen Making." *BG* 11 (1886): 132.
"Herbarium Cases." *BG* 11 (1886): 186–87.
"Books of Reference." *BG* 11 (1886): 193.
"Grasses for the Nebraska Plains." *Amer Agr* 45 (1886): 400.
"Clover upon the Nebraska Plains." *Amer Agr* 45 (1886): 443.
"How Nebraska Conducted Her State Fair." *Amer Agr* 45 (1886): 472.
"Destruction of Rocky Mountain Forests." *Amer Agr* 45 (1886): 520.
"Forage Plants." *Breeder's Gazette* (1886), 381.
"List of Nectar-Producing Plants." *Neb Hort Soc* (1887–88), 203–9.
"A Meeting-Place for Two Floras." *Bulletin of the Torrey Club* 4 (1887): 189–91.
"The Relations of Vegetation to Stock Growing." *Breeder's Gazette* (1887), 251.
"Tree Planting on the Plains." *Amer Agr* 46 (1887): 9.
"Trees and Blizzards." *Amer Agr* 46 (1887): 47.
"Horticulture in Nebraska." *Amer Agr* 46 (1887): 99.
"Ornamental Wild Flowers of the Plains." *Amer Agr* 46 (1887): 156.
"Some Western Weeds." *Amer Agr* 46 (1887): 199.
"The Country School and the Farmer's Boy." *Amer Agr* 46 (1887): 243.
"Arbor Day in Nebraska." *Amer Agr* 46 (1887): 327.

"A Good Western Grass." *Amer Agr* 46 (1887): 367.
"Two Promising Native Cherries." *Amer Agr* 46 (1887): 418.
"Agricultural Progress in Nebraska." *Amer Agr* 46 (1887): 481.
"A Model Agricultural Fair." *Amer Agr* 46 (1887): 551.
"Botanical Journals." *AN* 21 (1887): 79.
"The Study of Plant Diseases." *AN* 21 (1887): 276–77.
"Botanical Manuals for Students." *AN* 21 (1887): 376–79.
"The Growth of *Tulostoma mammosum*." *AN* 21 (1887): 665–66.
"The Eastward Extension of *Pinus ponderosa* Douglas, var. scopulorum." *AN* 21 (1887): 928–29.
"The Westward Extension of the Black Walnut." *AN* 21 (1887): 929.
"The Iron-Weed Tree in the Black Hills." *AN* 21 (1887): 929.
"A Duty Which We Owe to Science." *SPAS* (1887), 28–30.
"The Grass Flora of the Nebraska Plains." *AN* 22 (1888): 171–72.
"An Overlooked Function of Many Fruits." *AN* 22 (1888): 531.
"Elementary Agricultural Schools." *Amer Agr* 47 (1888): 14.
"What One Farmers' Club Did." *Amer Agr* 47 (1888): 49.
"Horticulture on the Plains." *Amer Agr* 47 (1888): 98.
"Fine Grasses for Fine Stock." *Breeder's Gazette* (1888), 212–13.
(With Herbert J. Webber.) "The Grasses and Forage Plants, and the Catalogue of Plants." *Neb Bd Agr* (1889), 144–74.
(With Webber.) "The Smut of Indian Corn." *Neb Exp St,* Bulletin #11 (1889), 295–305.
"Report of the Botanist." *Neb Bd Agr* (1889), 37–40.
"Wheat Rust." *Neb Exp St,* Press Bulletin #2 (1889).
"Education for the Farmer's Boy." *Nebraska Farmer* 13 (1889): 814, 847, 871, 883, 910.
"Two Big-Rooted Plants of the Plains." *AN* 23 (1889): 174–76.
"The Flora of the Upper Niobrara." *AN* 23 (1889): 537–38.
"The Grass Problem in Nebraska." *SPAS* (1889), 17–19.
"A Dozen Grasses and Clovers for Nebraska." *Neb Bd Agr* (1890), 100–8.
"Abuse of Cold Storage." *Breeder's Gazette* (1890), 8.
"The Native Grasses of the Plains." *Breeder's Gazette* (1890), 171.
"Grasses of the West Again." *Breeder's Gazette* (1890), 252.
"Wild or Tame Grasses for the Plains?" *Breeder's Gazette* (1890), 375.
"The Meadow in Winter." *Breeder's Gazette* (1890), 484–85.
"Some Useful Nebraska Grasses." *Amer Agr* 49 (1890): 635.
"A New Work on 'Plant Morphology.' " *AN* 24 (1890): 1082–83.
"The Completion of Saccardo's Sylloge Fungroum." *AN* 24 (1890): 675–76.
"The Host-Index of the Fungi of the United States." *AN* 24 (1890): 1196–97.
"Some Bad Station Botany." *AN* 24 (1890): 1197.
"Wheat Smut." *AN* 24 (1890): 1197–98.
"The Diseases of Farm and Garden Crops." *Nebraska Farmer* 14 (1890): 89, 129–30, 151, 165, 189, 209, 250, 293, 333.
"The Forage Problem on the Plains." *SPAS* (1890), 17–20.

"A Preliminary List of the Grasses of Nebraska." *Neb Bd Agr* (1891), 124–30.
"On the Fertilization, Crossing, and Hybridization of Plants." *Neb Hort Soc* (1891), 100–13.
"Report upon the Native Trees and Shrubs of Nebraska." *Neb Exp St,* Bulletin #18 (1891).
"Forage Plants on the Plains." *Breeder's Gazette* (1891), 468.
"The Bearberry in Central Nebraska." *AN* 25 (1891): 1130.
*Elementary Botanical Exercises for Public Schools and Private Study.* Lincoln: J. H. Miller, 1892.
"A Preliminary Report upon the Native Trees and Shrubs of Nebraska." *Neb Exp St* (1892), 171–98.
"A Second Report upon the Native Trees and Shrubs of Nebraska." *Neb Hort Soc* (1892), 154–85.
"Supplementary List of Recently Reported Species." *Neb Bd Agr* (1892), 45–53.
"A Preliminary Description of the Native and Introduced Grasses of Nebraska." *Neb Bd Agr* (1892), 209–79.
(With Albert Woods.) "Transpiration; or, the Loss of Water from Plants." *Contributions of the Botany Department* (1892).
"The Improvement of Wild Grasses." *Amer Agr* 51 (1892): 356.
"Notes on the Flora of the Black Hills." *Neb Acad Sci* 2 (1892): 17–19.
"The Weeds of Nebraska." *SPAS* (1893), 33–44.
"The Russian Thistle in Nebraska." *Neb Exp St,* Bulletin #31 (1893), 67–77.
"Evolution and Classification." *Proceedings of the AAAS* 42 (1893): 237–51.
"Simplification and Degeneration of Structure in the Angiosperms." *BG* 19 (1894): 369.
"Further Studies in the Relationship and Arrangement of the Families of Flowering Plants." *BG* 19 (1894): 372–73.
"The Homologies of the Uredineae (The Rusts)." *AN* 28 (1894): 989–96.
"The Botany of the Apple Tree." *Neb Hort Soc* (1894), 7–36.
"Some Facts in Vegetable Physiology Related to Problems in Irrigation." *Neb Bd Agr* (1894), 128–32.
"A Synopsis of the Larger Groups of the Vegetable Kingdom." *AN* 28 (1894): 63–65.
"The Russian Thistle in Nebraska." *SPAS* (1894), 286.
"Were the Sand Hills of Nebraska Formerly Covered with Forests?" *Neb Acad Sci* 5 (1894): 7.
"A Preliminary List of Honey-Producing Plants of Nebraska." *Neb Exp St,* Bulletin #40 (1895), 141–52.
"Sargent's Studies on the Forests of Japan." *AN* 29 (1895): 1049–56.
"Botany of the Grape." *Neb Hort Soc* (1895), 7–26.
"Progress of the Botanical Survey of Nebraska." *AN* 29 (1895): 580.
"A Protest Against the 'Rochester Rules.' " *AN* 29 (1895): 666–68.
"Notes on the Distribution of the Yellow Pine in Nebraska." *Garden and Forest* 8 (1895): 102–3.
"The Point of Divergence of Monocotyledons and Dicotyledons." *BG* 22 (1896): 229–32.

"Science and Culture." *Science* 4 (1896): 121–24.
"Grasses and Other Forage Plants." *Neb Bd Agr* (1896), 80–93.
"Notes on the Botany of the Strawberry." *Neb Hort Soc* (1896), 237–40.
"The Condition of Forests and Forestry: Nebraska." *Proceedings of the American Forestry Association* 2 (1896): 83–89.
"The Box-Elder on the Plains." *Garden and Forest* 9 (1896): 33.
"The Origin of the Flora of Nebraska." *Neb Acad Sci* 5 (1896): 11, 33.
"Phylogeny and Taxonomy of the Angiosperms." *BG* 24 (1897): 145–78.
"The Systematic Arrangement of the Protophyta." *AN* 31 (1897): 63–65.
"The Forage Problem in Eastern Nebraska, Central Nebraska, and on the High Plains." *Neb Bd Agr* (1897), 111–20.
"Trees on the Nebraska Plains—Are They Increasing or Diminishing from Natural Causes?" *Neb Bd Agr* (1897), 313–16.
"Are the Trees Receding from the Nebraska Plains?" *Garden and Forest* 10 (1897): 456–57.
"Report of the Botanist: A Preliminary Account of Diseases of the Farm Crops of Nebraska." *Neb Bd Agr* (1898), 129–61.
"Some Facts in Plant Physiology Bearing upon Horticultural Practices." *Neb Hort Soc* (1898), 135–45.
"Some Characteristics of the Foothill Vegetation of Western Nebraska." *AN* 32 (1898): 111–13.
"High School Botany." *Science* 7 (1898): 266–67.
"The Conference of Science Teachers in the Trans-Mississippi Educational Convention." *Science* 8 (1898): 123–26.
"Physiology of the Apple Tree." *Neb Hort Soc* (1899), 27–45.
"Relative Infrequence of Fungi upon the Trans-Missouri Plains and the Adjacent Foothills of the Rocky Mountain Region." *AN* 33 (1899): 215.
"The Passing of the Russian Thistle." *SPAS* (1899), 83–85.
"Observations on Buffalo Grass." *SPAS* (1899), 105–6.
"The Natural Spreading of Timber Areas." *Forester* 6 (1900): 240–43.
"The Modern Conception of the Structure and Classification of Diatoms." *AMS* 21 (1900): 61–85.
"The Modern Conception of the Structure and Classification of Desmids." *AMS* 22 (1901): 89–97.
"Notes on the Apple Scab." *Neb Hort Soc* (1901), 100–8.
"Twenty Native Forest Trees of Nebraska." *Forester* 7 (1901): 314–17.
"Old World Contributions to Western Orchards." *SPAS* (1901), 26–34.
"The Structure and Classification of the Conjugatae." *AMS* 23 (1902): 145–50.
"The Preservation of Wild Flowers." *Science* 16 (1902): 476.
"Evolution in Microscopic Plants." *AMS* 24 (1903): 5–12.
"The Structure and Classification of the Phycomycetes." *AMS* 24 (1903): 27–54.
"Preliminary Paper on Diseases of Grapes in Nebraska." *Neb Hort Soc* (1903), 86–89.
"Distribution of Native Forest Trees of Nebraska." *Neb Bd Agr* (1903), 257–80.
"Some Foreign Botanical Gardens and Parks." *Neb Bd Agr* (1903), 82–88. (also in *Neb Hort Soc* [1904], 157–66)

"Distribution of Forest Trees on the Nebraska Plains." *Atlantic Slope Naturalist* 1 (1903): 21.
"The Classification of Protophyta." *AMS* 25 (1904): 89–104.
"The Grasses of Nebraska." *Neb Bd Agr* (1904), 175–88.
"From the Crimea to Southern Poland." *TCF* (Mar. 2, 1904), 13.
"From Ararat to the Land of Golden Fleece." *TCF* (Apr. 6, 1904), 3, 15.
"Mountain Farming in the Caucasus." *TCF* (Apr. 27, 1904), 3, 15.
"Scientist's Notes on the Great Exposition [St. Louis]." *TCF* (Oct. 12, 1904), 3; (Oct. 19, 1904), 4; (Oct. 26, 1904), 9.
"Winnipeg and the Canadian Plains." *TCF* (Nov. 23, 1904), 7.
"Mountains and Glaciers in Canada." *TCF* (Dec. 14, 1904), 13.
"Notes on the Agriculture of the Caucasus Mountains." *SPAS* (1904), 37–43.
"The Structure and Classification of the Lower Green Algae." *AMS* 26 (1905): 121–36.
"Plant Migration Studies: Forest Trees." *University Studies,* 5 (1905): 11–37.
"Life in a Seaside Summer School." *Popular Science Monthly* 67 (1905): 80–89.
"How Much Plant Pathology Ought a Teacher of Botany to Know?" *Plant World* 8 (August 1905): 189–97.
"Has the Time Come When None but College Graduates Should Be Elected to High School Positions?" *UJ* 2 (December 1905): 1–2.
"The Wild Flower Preservation Society of America: Report of the President for 1904." *Plant World* 8 (February 1905): 39–42.
"Crop Improvement by Utilizing Wild Species." *Neb Hort Soc* (1906), 116–23.
"City Trees." *Neb Hort Soc* (1906): 153–56.
"The Forest Trees of Eastern Nebraska." *Ia Acad Sci* 13 (1906): 75–87.
"The Growing Importance of Plant Physiology in Agricultural Education." *SPAS* (1906), 50–54.
"A Synopsis of Plant Phyla." *University Studies* 7 (1907): 1–99.
"Field Work in Botany in Grammar and High Schools." *Nature-Study Review* 3 (January 1907): 9–16.
"A Word As to High School Botany." *UJ* 4 (October 1907): 11–12.
"The Physiology of Pruning." *Neb Hort Soc* (1908), 94–98.
*Outline of Plant Phyla.* Lincoln: University of Nebraska, 1909.
"Laying the Foundations." *Annals of Iowa* 9 (1909): 26–44.
"The Phyletic Idea in Taxonomy." *Science* 29 (1909): 91–100.
"A Plan for Completing the State Flora." *Science* 29 (1909): 594.
"The Phyla, Classes, and Orders of Plants." *AMS* 29 (1910): 85–96.
"Some European Forest Notes." *Forestry Quarterly* 8 (1910): 201–9.
"Laboratory Equipment for Botany in the High Schools." *UJ* 8 (1911): 35–39.
"The Content of High School Botany." *UJ* 8 (February 1912): 89–90.
"The Place for the Arts College." *UJ* 8 (May 1912): 132.
"Some of the Next Steps in Botanical Science." *Science* 37 (1913): 1–13.
"Tea Growing in Transcaucasia." *Pomona College Journal of Economic Botany* 3 (May 1913): 441–45.

"A Preliminary Paper on Drouth Endurance." *Neb Hort Soc* (1913), 245-48.

"Instruction in Pure Science for Agricultural Students." *Proceedings of the Association of American Agricultural Colleges and Experiment Stations* (1914), 213-18.

"Synopsis of the Conjugate Algae—Zygophyceae." *AMS* 33 (1914): 11-49.

"Revisions of Some Plant Phyla." *University Studies* 14 (1914): 1-73.

"The Phylogenetic Taxonomy of Flowering Plants." *Annals of the Missouri Botanical Garden* 2 (1915): 112-15.

# INDEX

Abbot, Theophilus C., 6
Abrams, LeRoy, 181
Adanson, Michel, 186
Affiliated Societies of Agricultural Science, 124
Agassiz, Louis, 25
Agricultural botany, 15–17, 40–41, 94, 97–99, 123
Agricultural colleges, nature of, 4–12, 40, 43–46, 47–55, 61–63, 71, 125
Agricultural education, 3–12, 18–20, 23, 40–41, 43–46, 47–55, 61–63, 167–68
Agricultural experiment stations, 20–21, 39, 54, 56–58, 70, 98–99
Agricultural newspapers and journals, 6, 9, 42–43, 52, 62, 70, 122
Agricultural science, 9, 15, 18–20, 23, 39–42, 47–71, 122–27, 167–68
Agriculture
 and pathology, 15–17, 58–60, 64–65, 68–69
 professionalization of, 52–55, 97–99, 177–78
Agriculturists
 and education, 37, 41, 64–69, 122, 126
 and science, 9–10, 45, 50–52, 61–63, 169
*American Agriculturist,* 6, 43, 52, 122
American Association for the Advancement of Science (AAAS), 13, 77–80, 94–99, 172
 and agriculture, 123–24
 Biology Section of, 78–79, 94, 96–97, 99, 100
 Botanical Club of, 78–79, 89, 94–97, 99, 100–103, 108, 110–12, 119, 148
 Botany Section of, 96, 100, 107–8, 110, 116, 119, 130
 nomenclature committee of, 103, 111, 112–13, 127
 and professionalization, 100, 115, 116–17, 119–20, 122
American Breeders' Association, 167
American Forestry Association, 150
*American Journal of Science,* 30, 94
American Microscopical Society, 97, 194, 199
American Mycological Society, 129
*American Naturalist,* 13, 30–31, 74–77, 93–94, 180
American Society of Agronomy, 124
American Society of Naturalists (ASN), 119–20, 128
Ames, Adeline, 176
Andrews, E. Benjamin, 154
Arber, F. A. Newell, 197
Arthur, Joseph C., 29, 36, 56
 and plant diseases, 38, 94
 and professionalization of botany, 72, 74–75, 77, 78–79, 95, 105, 115, 119
Association of American Agricultural Colleges and Experiment Stations (AAACES), 42, 58, 98–99, 123–24, 167
Atkinson, George F., 129
Aughey, Samuel, 131
Avery, Samuel, 178

Bailey, Liberty Hyde, 118, 125, 168–69, 181, 182–83, 184
Barnes, Charles R., 91, 128
 and professionalization of botany, 74–75, 77, 108, 115, 118, 119, 129, 130
Bary, Anton de, 27, 36, 89, 190
Bates, Carlos, 149
Bates, J. M., 135
Beal, William J., 42, 65, 72, 79, 91, 95, 98
Beattie, R. Kent, 114

255

Bentham, George, 27, 36, 134, 186–89, 195, 199
Bessey, Charles
  AAACES, 98–99
  AAAS, 13, 77–80, 94–97, 100–103, 105, 107, 119, 150, 172, 176, 179, 199
  as acting chancellor, 60–63, 77, 120
  and agricultural education, 8–9, 11, 40–41, 45, 47–55, 61–63, 71, 167–69, 177–78
  and agricultural science, 8–9, 15–16, 18, 21, 37–43, 47–63, 64–71, 97–99, 122–27, 167–68
  and agriculture, popularization of, 9–10, 37, 42–43, 51–52, 61–62, 66–70, 126–27, 167
  *American Naturalist,* 34, 36, 42, 74–77, 93–94, 97, 120–21
  American Society of Naturalists, 120
  Anti-Saloon League (Lincoln), 159
  and athletics, 176–77
  Botanical Club, 78–79, 94–95, 97, 100, 119, 148
  and botanical education, 24–34, 35, 40–41, 48–49, 63–64, 80, 92–93, 167–68, 171–75
  and botanical research, 13–17, 34, 38–39, 64–65, 67–68
  Botanical Society of America, 109, 117–19, 128, 192, 199
  and botanical survey of Iowa, 16–17, 36
  Botanical Survey of Nebraska, 65–66, 115, 131–40, 143, 147, 170–71
  *Botany for High Schools and Colleges,* 27–33, 63, 89
  Carnegie Institution, 161–62
  Civic Federation of Lincoln, 163
  and classification, 27, 28, 30–31, 35–36, 64, 90–91, 107, 117, 121, 134, 179, 185–99
  College of Literature, Science and Arts, 60, 61, 63
  Committee on Secondary School Studies, 171–72
  and conservation, 149–56, 160, 163–65, 170
  and Coulter, John M., 75–80, 93–94, 119, 171–75
  and Darwin, Charles, 35, 38, 73, 89, 147, 187
  Dry Farming Congress, 167
  and ecology, 133, 135, 136–37, 143–49, 156, 169–71, 173–74
  *Elementary Botanical Exercises,* 64, 172, 186, 192
  *Essentials of Botany,* 33–34, 63, 93, 192, 198
  and evolution, 35, 38, 91, 107, 179–80, 183–91, 193–99
  and flowering plants, 33, 131–32, 186, 190, 192, 194–99
  and forage plants and grasses, 52, 58, 65–67, 70, 133, 136–37
  and forest reserves, 149–53, 156, 163–65, 169
  and forestry and trees, 139–47, 149–56, 163–65, 169–70
  and German botany, 27, 28, 32, 37, 186, 188, 192
  and government, 159–66
  and grasslands, 136–40, 147
  and Gray, Asa, 12–15, 26–28, 30, 80, 89, 114, 167, 183, 187, 190
  Great Plains, 65–67, 133, 136, 138–47, 151–53, 156, 170
  Hatch Act, 56–58
  and heredity, 6, 180, 191
  Industrial College, 47–55, 57, 60–63, 68, 70–71, 154, 175–78
  Iowa Academy of Sciences, 21–22
  Iowa Agricultural College, 7–12, 20, 24, 40, 43–46
  Iowa State Horticultural Society, 18, 21, 37, 38–39
  and laboratory method, 14, 15, 24–25, 27, 29, 30, 31, 40, 48–49, 91–93, 132, 172
  Louisana Purchase Exposition, 171
  Madison Botanical Congress, 105, 107, 109
  Michigan Agricultural College, 3, 6–7, 11, 13, 25
  and nature study, 64, 172–73
  Nebraska Agricultural Experiment Station, 57–58, 62, 67, 68–70, 140, 167, 175
  Nebraska State Board of Agriculture, 47, 50, 52, 55, 58, 61–62, 68, 69,

# INDEX

131, 141, 159, 167
Nebraska State Horticultural Society, 52, 55, 62, 67–68, 69, 141, 167
and new botany, 31, 34, 47–49, 72–80, 89–97, 104, 107, 110, 118, 121, 147, 167–68, 185, 199
and nomenclature reform, 102, 106–7, 112–15
and nonflowering plants, 13–17, 26, 28, 30–31, 35, 36–38, 55, 58, 64–65, 68–69, 70, 76, 80, 89, 92–93, 113, 135, 192–94
North London Christian Evidence League, 183
and paleobotany, 196–98
and pathology, 15–17, 37–38, 55, 62, 64–65, 68–69, 79, 95, 109, 121, 168
and plant geography, 38, 91, 121, 133–39
and professionalization, 74–80, 93–99, 100–101, 107–10, 118–24, 128
and Progressivism, 157–66
and religion, 183
*Science,* 120–22
and science in society, 165–66
and scientific education, 161, 166
and secondary education, 33, 93, 171–75
Smithsonian Institution, 199
and social and moral issues, 158–60, 166
Society for Plant Morphology and Physiology, 120
SPAS, 42, 70, 98, 122–24
and structural physiology, 6–7, 13–15, 21, 28–35, 49, 54, 64, 67–68, 75–76, 80, 89–91, 93, 95, 97, 121, 145–49, 168, 173–74, 175, 186, 190–99
and systematic botany, 6, 13–17, 32–33, 34, 64, 76, 80, 90–91, 102, 106–7, 112–15, 131–36, 167–68, 175. *See also* Bessey and classification
and textbooks, in botany, 25–34, 35, 64, 80, 89–90, 180, 192, 194, 198
Trans-Mississippi Exposition, 171
USDA, 41, 49, 55–56, 69–70, 97–98, 134, 150–53, 154, 156, 162, 167, 169–70, 176
University of California, 14, 25
University of Minnesota, 25, 77, 79, 167
University of Nebraska, 46–49, 57–58, 60–63, 173, 175–78
Washington State College, 167
and Welch, Adonijah S., 7–12, 20, 25, 45
Wild Flower Preservation Society of America, 170
World's Industrial and Cotton Centennial Exposition, 47
Bessey, Ernst, 167, 176, 180
Bessey, Lucy Ahearn, 14
Billings, Frank S., 58–60, 63, 68
Blankenship, J. W., 145
*Botanical Gazette,* 30–31, 74–75, 77, 89, 92–98, 104–5, 110, 111, 121, 180, 188, 189
Botanical Society of America, 103–4, 108–9, 115–19, 128–30, 192, 199
Botanists of the Central States, 128, 129
Botany
 agricultural, 15–17, 40–41, 94, 97–99, 123
 applied, 15–20, 37, 54–55, 64–71, 97–98, 167–68
 classification, 27, 28, 36, 90, 186–99
 ecology, 173–74
 education, 6, 26–33, 40–41, 64, 89–90, 91–93, 94, 107, 167–68, 171–75
 embryology, 130, 190–91, 196
 evolution (adaptation, isolation, natural selection, organic selection, orthogenesis), 12, 26, 72–73, 130, 135, 179–99
 experimental, 73–74, 130, 147, 184–85, 187, 198
 flowering plants, 80, 89, 92, 196
 heredity, 179–85
 horticulture, 7, 18–21, 38–39, 47–48, 52, 54, 67–68
 journals of, 74–77, 89, 93–94, 96
 laboratories, development of, 13–15, 24–25, 30–31, 33, 34, 73, 91–93
 Madison Botanical Congress, 104–10
 morphology, 30, 34, 72–74, 90, 190–99

Botany (*continued*)
  new botany, development of, 72–80, 89–97, 103–4, 107, 110, 120, 128, 184–85, 186
  nomenclature reform
    American Code, 127–28
    Checklist of Flowering Plants, 103, 106, 108–15, 119, 121, 188
    Harvard Circular of Protest, 111–15
    international botanical congresses, 103, 104–5, 109, 127–28
    International Code of Nomenclature (Paris Code), 102, 105, 111, 115, 128
    International Committee on Nomenclature, 105
    Rochester Rules, 101–15, 116, 118, 119, 121, 127–28, 134, 135
  nonflowering plants, 15–17, 28–31, 36, 64–65, 68–69, 89
  paleobotany, 196
  pathology, 15–17, 37–38, 72, 79, 110
  physiology, 20, 27, 30–34, 72–74, 75–76, 89–90, 95, 97
  plant geography, 73–74, 76, 94, 137–38
  professionalization, 72–80, 93–99, 100–101, 105, 107–10, 114–22, 127–30, 184–85
  systematic, 20, 27, 34, 72–74, 76, 89–91, 101–8, 111–15, 127–28, 186–91, 199
  textbooks, 13, 25–34, 80, 89–90, 92–93, 173, 189
Brewer, W. H., 42
Britton, Elizabeth, 170
Britton, Nathaniel L., 101, 114, 116, 121–22, 135, 186, 188
  and Addison Brown, *An Illustrated Flora*, 113, 115, 188
  and nomenclature reform, 102–3, 105, 107, 112–13, 115, 127
Brown, Addison, 113, 115, 188
Bruner, Lawrence, 140
Bryan, William Jennings, 149, 165
Buchanan, James, 5
Budd, J. L., 18, 39, 40, 53
*Bulletin of the Torrey Club*, 74, 110, 115
Burrill, Thomas J., 36, 38, 79, 91

Caldwell, George C., 42
Caldwell, Otis W., 174
Campbell, Douglas H., 91, 93, 171, 186, 188
Candolle, Alphonse de, 65, 107, 134
Candolle, Augustin de, 186–88, 190
Canfield, James, 63
Cannon, Joseph, 163
Carpenter, C. C., 56
Cattell, J. McKeen, 121–22
Chapman, A. W., 80
Clements, Frederic E., 133–35, 138, 148–49, 156, 169–70, 174, 175
Cleveland, Grover, 149
Commission on Country Life, 168–69
Condra, George, 169
Cook, O. F., 181
Cooke, M. C., 17, 80
Cope, Edward Drinker, 75, 77, 120–21, 180, 181
Coulter, John M., 137, 196
  and education, 171–75
  and new botany, 28, 31, 34, 72, 74–75, 77–80, 90–91, 92–94
  and nomenclature, 103, 105, 115
  and professionalization, 77–80, 116, 119, 122
Coulter, Stanley, 91
Coville, Frederick, 103, 112–13, 134, 188
Cronquist, Arthur, 199
Curtiss, A. H., 49

Darwin, Charles (Darwinism), 20, 35, 38, 73–74, 89, 147, 184, 187
Dawson, J. W., 118, 181
Dietrich, Charles H., 154
Drude, Oscar, 137–38
Dudley, W. R., 118

Eaton, Daniel, 116–17
Edgerton, C. W., 155
Education. *See* Agricultural education; Bessey and botanical education; Botany, education
Eichler, August W., 186, 188
Elliot, Charles W., 171
Ellis, J. B., 113
Emerson, R. A., 184
Englemann, George, 36, 196

# INDEX

Engler, Adolf, 36, 138, 188–91, 193, 195, 197, 198, 199

Farlow, William G., 27, 29
 and fungi, 14, 17, 36, 89
 and professionalization, 72, 78, 79, 116–17, 118, 119
Fernald, Merritt, 127, 189
Fernow, Bernhard E., 140–43, 150, 152
Field, Ethel, 176
Fleming, Ruth Bates, 176
Forestry. *See* Bessey, forest reserves, forestry and trees; USDA, Bureau and Division of Forestry
Furnas, Robert W., 47, 52, 61

Galloway, Beverly T., 129
Ganong, William F., 106, 128, 130, 173–74
Garfield, James, 163
Geddes, James L., 44, 46
Gere, Charles H., 59
Gill, Theodore N., 181
Gilman, Daniel C., 14
Goebel, Karl, 188, 190, 192
Goodale, George L., 13, 25–26, 27, 29, 34, 89–90, 95
Grant, Ulysses S., 160
Grasslands, 133–47
Gray, Asa, 12–15, 16, 36, 74, 95, 183, 187, 189, 190
 *Darwiniana,* 183
 *Manual,* 65, 90, 103, 114, 167, 189, 193
 and nomenclature, 101–2, 107, 114–15
 and textbooks, 26–31, 33, 64, 80, 90, 167
Great Plains, 64, 65–67, 133–36, 138–47, 151–53, 156, 170
Greene, Edward, 101–2, 105, 112, 114
Grisebach, A. H. R., 138

Haeckel, Ernst, 186
Hall, William L., 151–53
Hallier, Hans, 188
Halsted, Byron D., 36, 180, 187
Harrison, Carrie, 176
Harvard College, 13–14, 91
Hatch Act, 56–58

Hicks, Lewis E., 60, 142
Hilgard, Eugene W., 30
Hofmeister, Wilhelm, 27, 190
Hoggett, Lucien, 10
Holm, Theodore, 113
Holt, Henry and Company, 25–26, 30, 32–33, 64
Holt County (Nebraska) Experimental Tree Plot, 140–41, 142, 151, 152
Hooker, Joseph, 27, 36, 101–2, 134, 186–89, 195, 199
Horticulture. *See* Botany, horticulture
Hutchison, John, 199
Huxley, Thomas, 20

Iowa
 Academy of Sciences, 21–22
 Grange, 10, 45
 State Agricultural Society, 20, 39–41
 state entomologist, 20
 State Horticultural Society, 18–21, 37–39, 44
 State University of Iowa, 32
Iowa Agricultural College
 beginnings of, 7–12, 23–25
 botanical survey, 16–17, 36
 and disputes over land-grant education, 7–12, 18–20, 40–46
 and Farmers' Institutes, 9
 and manual labor education, 11, 23–24
 and State Horticultural Society, 18–21, 39
 and Welch group, 8–12, 18–20, 23, 39, 43–46, 47

James, Thomas, 36
Jones, Marcus, 114
Jordan, David Starr, 181
Jordan, W. H., 126
*Journal of Agricultural Sciences,* 123
*Journal of Mycology,* 89, 94
Jussieu, Antoine-Laurent de, 186, 187

Kellerman, William A., 103, 118
Kew Gardens, 101, 104, 114
Kingsley, John S., 143, 181–82
Knapp, Seaman A., 39–40, 43–46
 Knapp-Carpenter bill, 56
Kuntze, Otto, 106, 108

Lamarckism, 179–85
Lamson-Scribner, Frank, 55, 118, 134
Land-grant movement, 4–12, 40, 43–46, 47–55, 61–63, 71, 125. *See also* Bessey and agricultural education
Lemmon, J. G., 49
Lesquereux, Leo, 36
Lindley, John, 186
Linnaeus, Carl, 104, 107, 186, 187
Lloyd, Francis E., 181
Loring, George B., 41, 49, 56

McAfee, H. H., 18–20, 53, 139
MacDougal, Daniel T., 130, 184
Macfarlane, John, 181
McKinley, William, 149
MacMillan, Conway, 107, 115, 188
McNab, William Ramsey, 27, 32
Macomber, J. K., 139
Manatt, Irving J., 47, 59–60
Marsland, Herbert, 133
Mendelism, 184
Michigan Agricultural College, 3, 6–7, 8, 11, 13, 25, 200
Miles, Manly, 6
Millard, J. H., 162
Minnesota, University of, 25, 77, 79, 167
Minnesota botanical survey, 115
Missouri Botanical Garden (Shaw Garden), 91, 133, 176, 198
Mohl, Hugo von, 190
Morrill, Justin S., 5
Morrill Act, 5, 7, 12, 19, 40, 46, 47, 49, 51, 61
Muir, John, 167

Nageli, Karl Wilhelm, 27, 190
National Arboretum, 95
National Education Association (Committee of Ten), 171–72, 173
National Herbarium, 95, 109, 161
Natural history, 25, 29, 30, 48, 171
Nebraska
 Academy of Sciences, 120, 153
 Conservation Commission of Nebraska, 164
 forest reserves, 153–54, 156
 livestock associations, 52, 59
 *Nebraska Farmer*, 62, 69

*Phytogeography of Nebraska*, 138, 148
Rural Life Commission, 168–69
sandhills, 66, 131, 133, 134, 135, 138, 140–44, 151–54, 156
State Agricultural Society, 142
State Board of Agriculture, 47, 50, 52, 61–62, 134
State Horticultural Society, 62, 120, 134, 152
State Park and Forest Association, 150
Nebraska, University of
 Agricultural Experiment Station, 57–58, 59–60, 62, 66, 67, 68–70, 134, 175, 177
 Botanical Survey, 65–66, 115, 133–39, 140, 143, 147, 170–71
 College of Agriculture, 49–50, 177
 College of Literature, Science and the Arts, 60, 61, 63, 177–78
 Department of Botany, 48–49, 57–58, 68–69, 149, 167, 175–77
 Department of Forestry, 154–56
 disputes over agricultural program, 50–52, 55, 57–60, 61–63, 177–78
 Farmers' Institutes, 51–52, 61–62, 69
 Industrial College, 47–55, 57, 59–63, 68, 70–71, 154–56, 175–78
 Patho-Biological Laboratory, 57, 58–60, 70
 Sem Bot, 82, 132–37
 University Alpine Station, 149, 167
Nelson, Aven, 114
Neo-Darwinism, 180–85
Neo-Lamarckism, 179–85
New Botany. *See* Botany
Newcomb, Simon, 124
Newcombe, F. C., 174
Newell, F. H., 152

Outdoor Art League of California, 170

Packard, Alpheus S., Jr., 29, 31, 75, 77, 180
Palmer, Charles S., 172
Pammel, L. H., 95
Parkin, John, 197
Parry, C. C., 19, 38, 49
Peck, Charles H., 17, 79, 113

# INDEX

Penhallow, D. P., 188
Phelps, Almira Hart Lincoln, 27–28
Pinchot, Gifford, 150, 152–53, 155, 169–70
Pool, Raymond J., 24, 149, 156
Pound, Roscoe, 132–34, 137–38, 148
Prairies, 64, 136–40, 144–47, 150
Prantl, Karl, 89, 188–91, 193
Prentiss, Albert N., 6, 91
Progressivism, 157–66

Rebmann, Jeremiah, 154–55
Riley, C. V., 21
Roberts, Isaac, 10
Robinson, Benjamin L., 101, 102, 106, 111–12, 116–17, 119, 189
Roosevelt, Theodore, 153, 160, 165, 168
Rothrock, Joseph T., 34, 36, 72, 73, 79, 91
Rowland, Henry A., 124
Rowless, W. W., 118
Rusby, H. H., 103
Rydberg, Pier A., 133–36

Sachs, Julius von, 27–28, 30–31, 32, 36, 75, 89, 190
Salmon, D. E., 59
Sargent, Charles S., 117, 150, 167
Saunders, DeAlton, 136
Savage, Ezra P., 152, 153
Schwendener, Simon, 27
*Science,* 94, 120–22, 180
Scott, Charles A., 154
Sem Bot, 82, 132–37
Setchell, William A., 118
Seymour, A. B., 89
Shear, C. L., 134–36, 176
Smith, Erwin F., 106, 114, 118, 121, 125, 176
Smith, Jared, 70, 132–33
Smith, John Donnell, 118
Society for Horticultural Science, 124, 129
Society for Plant Morphology and Physiology, 120, 128, 129–30, 173
Society for the Promotion of Agricultural Science (SPAS), 41–42, 97, 98–99, 122–24
Society of American Bacteriologists, 129
Spalding, Volney M., 91

Spencer, Herbert, 20, 38
Spillman, W. J., 181
Stakler, Millikan, 10
Steele, J. Dorman, 28
Strasburger, Eduard, 27, 190
Sturtevant, E. Lewis, 41–42, 56, 65
Sudworth, George B., 170

Taft, William Howard, 163, 165
Takhtajan, Armen, 199
Tansley, A. G., 193
Teller, Henry, 163
Thaxter, Roland, 116–17, 118, 119
Thayer, John, 59
Thiel, Albert F., 170
Thompson, S. R., 50
Thurber, George, 6
Tippo, Oswald, 199
Toliver, J. C., 141–42, 153
Torrey, John, 16
Trelease, William, 72, 91, 109, 116–18, 122
True, Alfred C., 130
Tuckerman, Edward, 36
Turner, Jonathan B., 5
Tyndall, John, 20, 124

Underwood, Lucien, 36, 72, 103, 104–5, 114–15
United States Department of Agriculture (USDA), 21, 41, 55–56, 97–98, 115, 162, 176
  Bureau of Animal Industry, 58–60
  Bureau of Forestry, 153–55, 169–70
  Bureau of Plant Industry, 70, 170, 176
  Division of Botany, 49, 55, 134
  Division of Forestry, 140–41, 150, 151–52
  Division of Vegetable Physiology and Pathology, 69–70, 176
  Section of Tree Planting, 151–53
  Section of Vegetable Pathology, 55

Vasey, George, 65, 109
Vries, Hugo de, 183–84

Walker, Elda Rema, 175
Wallace, Alfred Russel, 38
Ward, Lester Frank, 103, 118, 181, 186, 196

Warming, Eugen, 188
Watson, Sereno, 36, 80, 102
Weaver, J. E., 170
Webber, Herbert, 65, 132–33, 176, 181, 184
Weismann, August, 182–83
Welch, Adonijah Strong, 7–12, 18–20, 23, 25, 39, 43–46
Western Society of Naturalists, 93, 97
White, David, 196
Wieland, G. R., 197
Williams, Thomas A., 132–33, 135–36
Wilson, Edmund B., 95
Wilson, William D., 10
Wing, H. H., 52, 62
Wood, Alphonso, 27–28, 31
Woods, Albert F., 70, 97, 114, 132–33, 167, 176, 199
Wynn, W. H., 12

Yale Forest School, 154–55

## DATE DUE

DUE DATE SUBJECT TO CHANGE
IF A RECALL IS REQUESTED

Demco, Inc. 38-293